Sequential Decision Analytics and Modeling: Modeling with Python

Warren B. Powell
Professor Emeritus
Princeton University
wbpowell328@gmail.com

the essence of knowledge
Boston — Delft

Foundations and Trends® in Technology, Information and Operations Management

Published, sold and distributed by:
now Publishers Inc.
PO Box 1024
Hanover, MA 02339
United States
Tel. +1-781-985-4510
www.nowpublishers.com
sales@nowpublishers.com

Outside North America:
now Publishers Inc.
PO Box 179
2600 AD Delft
The Netherlands
Tel. +31-6-51115274

The preferred citation for this publication is

W. B. Powell. *Sequential Decision Analytics and Modeling: Modeling with Python.* Foundations and Trends® in Technology, Information and Operations Management, vol. 15, no. 4 and vol. 16, no. 1-2, pp. 1–313, 2022.

ISBN: 978-1-63828-082-8

Contents

Sequential Decision Analytics and Modeling: Modeling with Python

Warren Powell

Professor Emeritus, Princeton University, USA; wbpowell328@gmail.com

---

ABSTRACT

Sequential decision problems arise in virtually every human process, spanning finance, energy, transportation, health, e-commerce and supply chains. They include pure learning problems as might arise in laboratory (or field) experiments. It even covers search algorithms to maximize uncertain functions. An important dimension of every problem setting is the need to make decisions in the presence of different forms of uncertainty, and evolving information processes. This book uses a teach-by-example style to illustrate a modeling framework that can represent *any* sequential decision problem. A major challenge is, then, designing methods (called policies) for making decisions. We describe four classes of policies that are universal, in that they span any method that might be used, whether from the academic literature or heuristics used in practice. While this does not mean that we can immediately solve any problem, the framework helps us avoid the tendency in the academic literature of focusing on narrow classes of methods.

Preface

My work in sequential decision problems grew out of research that started in the 1980s in trucking, and over my career spanned rail, energy, health, finance, e-commerce, supply chain management, and even learning for materials science. Sequential decision problems arise in daily activities such as sports, cooking, shopping, and finding the best path to a destination. They also arise when designing a product for a startup, hiring people for the startup, and designing marketing campaigns.

The early work in sequential decision problems (known as dynamic programs or optimal control problems) focused on solving a famous, and famously intractable, equation known as Bellman's equation (or Hamilton-Jacobi equations for continuous problems). I joined a community that worked on methods for approximating these equations; this work resulted in a successful book on approximate dynamic programming, producing a breakthrough for a class of resource allocation problems. Over time, however, I came to realize that approximate dynamic programming was a powerful method for solving a very narrow range of problems - the proverbial hammer looking for a nail.

My work on a variety of problems made me realize the importance of using a broad range of methods which could be found through the research literature. I found I could model any sequential decision problem with the same framework which involved searching over methods for making decisions, generally known as "policies" in the research literature. I was then able to organize the vast range of methods into four broad classes (meta-classes) of policies which span *any* method for making decisions, including anything proposed in the literature or used in practice (including methods that have not been invented yet!).

This framework is the foundation of a graduate-level book that I finished in 2022 called *Reinforcement Learning and Stochastic Optimization: A unified framework for sequential decisions.*[1] As I was writing this book, I realized that sequential decision problems are universal, arising in every human activity. Furthermore, these ideas could (and should) be taught to a broad audience, and not just the typical, analytically sophisticated crowd that we find in operations research, computer science, economics, and pockets of engineering.

The goal of this book is to enable readers to understand how to approach, model and solve a sequential decision problem, even if they are never going to write a line of code. While this book is analytical, the real goal is to teach readers how to *think* about sequential decision problems, breaking them down into the five core elements of a sequential decision model, modeling uncertainty, and then designing policies.

Just as there are many styles for teaching statistics within different communities, I believe there will be a similar evolution to teaching these ideas to different audiences. The examples in this book come from operations research, which I like to call the mathematics of everyday life. I think readers will find most of the examples to be familiar, independent of their professional field. At the same time, I can easily see versions of the book designed purely for different problem domains such as health, finance, energy, robotics, and supply chain management (and this is hardly a comprehensive list).

[1](Powell, 2022)

1

Modeling Sequential Decision Problems

The process of solving any physical problem (and in particular any sequential decision problem) on the computer requires building a mathematical model, as illustrated in Figure 1.1. For decades, the research community has used a standard mathematical framework for decision problems where all the data is known in advance (known as deterministic optimization). A simple version of a deterministic optimization problem, known as a linear program, might be written

$$\min_x c^T x, \tag{1.1}$$

where x is a vector of elements that have to satisfy a set of constraints which are typically written

$$\begin{aligned} Ax &= b, & (1.2)\\ x &\geq 0. & (1.3)\end{aligned}$$

It is not necessary to understand Equations (1.1)-(1.3) (which requires basic familiarity with linear algebra), but thousands of students graduate each year from courses where they learn this notation, and also learn how to translate a wide range of physical problems into this notation. Then, there are software packages that translate problems in

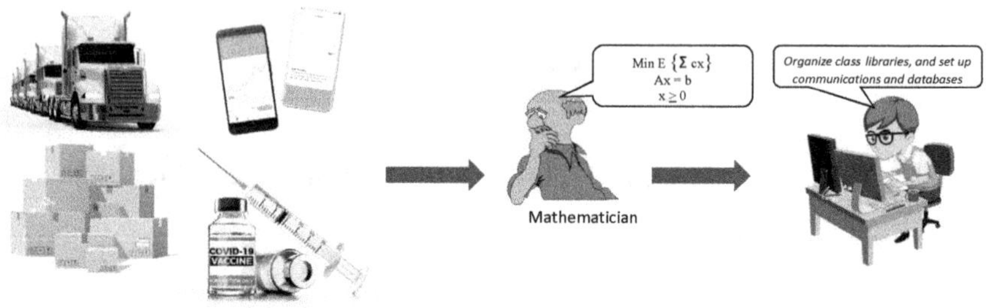

Figure 1.1: The bridge between the real world and the computer is a mathematical model.

this format into a solution. Most important, this notational language is spoken around the world. The same statement can be made about statistical modeling/machine learning which today is a much larger community than the people who understand Equations (1.1)-(1.3).

We cannot make the same statement about sequential decision problems which is a problem class that is studied by at least 15 different communities using eight fundamentally different notational styles, often using mathematics that requires advanced training. In this book, we use a teach-by-example style to show how to model the incredibly rich class of problems that we call sequential decision problems. While we focus on relatively simpler problems, our framework can be used to model *any* sequential decision problem. In addition, the resulting model can be translated directly to software.

The analytical foundation of this book is contained in Powell (2022) (RLSO), which is a graduate level text centered on methodology. From time to time we will refer to material in this book for readers who might be interested in greater depth, and we encourage technically inclined readers to use RLSO as a reference. However, it is not needed. This book is designed to provide the contextual background in the form of a series of examples that should enable readers to think clearly and precisely about sequential decision problems, even if they will never write a line of code.

This book is aimed at undergraduate or masters level students who have taken a course in probability and statistics (a knowledge of linear programming is not necessary, although we have an example which requires solving a linear program). All the chapters are built around specific examples, with the exception of Chapter 1, which provides an overview of the entire modeling framework, and Chapter 7, where we pause and use the first six chapters to illustrate some important principles.

The presentation should not require mathematics beyond what would be expected in a first course on probability and statistics. This said, the book is centered on showing how to describe sequential decision problems using notation that is precise enough that it can be the basis of computer software.

Python modules accompany most of the chapters; these modules were written around the modeling framework that runs throughout the book. At the same time, any software package that simulates a sequential decision problem, regardless of how it is being solved, can be translated directly into the modeling framework we use. For this reason, we encourage readers to look at any piece of notation as a variable in a computer program.

1.1 Getting started

Sequential decision problems can always be written as

$$decision, information, decision, information, decision, \ldots$$

Each time we make a decision we incur a cost or receive a contribution or reward (there are many ways to measure performance). Decisions are made with a method that we are going to refer to as a *policy*. A major goal that is a central focus of this book is to design effective policies that work well over time, in the presence of the uncertainty of information that has not yet arrived.

Sequential decision problems are ubiquitous, arising in virtually every human process. Table 1.1 provides a sample list of fields, with examples of some of the decisions that might arise. Most of these fields probably have many different types of decisions, ranging in complexity from when to sell an asset or to adopt a new web design, to choosing the best drug, material, or facility to design, to managing complex supply chains or dispatching a fleet of trucks.

Even more challenging than listing all the types of decisions is identifying the different sources of uncertainty that arise in many applications. Human behavior, markets, physical processes, transportation networks, energy systems and the broad array of uncertainties that arise in health hint at the diversity of different sources of uncertainty.

As this book is being written, humanity is struggling with the spread of variations of COVID-19. Dealing with this pandemic has been described as "mind-bogglingly complex," [USA Today, Sept 8, 2020] but this is really a byproduct of a failure to think about the problem in a structured way. We are going to show the reader how to break down problems into a series of basic components that lead to practical solutions.

Our approach starts by identifying some core elements such as performance metrics, decisions, and sources of uncertainty, which then lead to the creation of a mathematical model of the problem. The next step is usually (but not always) to implement the model on the computer, but there are going to be many problems where the process of building a computer model is impractical for any of a number of reasons. For this reason, we are also going to consider problems where we have to test and evaluate ideas in the field. To improve performance, we need to first learn how to make good decisions over time (this is how we control the system). Then, we turn to the design of the system.

At this time, the academic community has not adopted a standard modeling process for sequential decision problems. This is in sharp contrast with the arena of static, deterministic optimization problems which have followed a strict framework since the 1950s (Equations (1.1) - (1.3) represent a sample of this framework). Our modeling process is based on the presentation in Powell (2022), which is a book aimed at a technical audience that is primarily interested in developing and implementing models on the computer.

By contrast, this book is aimed at a broader audience that is first and foremost interested in learning how to *think* about sequential decision problems. It uses a teach-by-example style that focuses on communicating the modeling process which we feel can be useful even without

Table 1.1: A sample of different fields and choices that need to be made within each field (adapted from Powell, 2022).

Field	Questions
Business	What products should we sell, with what features? Which supplies should you use? What price should you charge?
Economics	What interest rate should the Federal Reserve charge given the state of the economy? What levels of market liquidity should be provided?
Finance	What stocks should a portfolio invest in? How should a trader hedge a contract for potential downside?
Internet	What ads should we display to maximize ad-clicks? Which movies attract the most attention? When/how should mass notices be sent?
Engineering	How to design devices from aerosol cans to electric vehicles, bridges to transportation systems, transistors to computers?
Public health	How should we run testing to estimate the progression of a disease? How should vaccines be allocated? Which population groups should be targeted?
Medical research	What molecular configuration will produce the drug which kills the most cancer cells? What set of steps are required to produce single-walled nanotubes?
Supply chain management	When should we place an order for inventory from China? Which supplier should be used?
Freight transportation	Which driver should move a load? What loads should a truckload carrier commit to move? Where should drivers be domiciled?
Information collection	Where should we send a drone to collect information on wildfires or invasive species? What drug should we test to combat a disease?
Multiagent systems	How should a large company in an oligopolistic market bid on contracts, anticipating the response of its competitors?
Algorithms	What stepsize rule should we use in a search algorithm? How do we determine the next point to evaluate an expensive function?

ultimately creating computer models. Central to our approach is the creation of a mathematical model that eliminates the ambiguity when describing problems in plain English. For readers who are interested in developing computer models, notation is the stepping stone to writing software. However, we are going to primarily use mathematical notation to create clarity when describing a problem, even if the reader never intends to write a line of code.

Our presentation proceeds as follows:

- Chapter 1 provides a light introduction to the universal modeling framework, illustrated using two inventory problems (a simple one, and one that is slightly more complex), followed by a brief discussion of modeling uncertainty. It then provides an introduction to the four classes of policies that cover every method for making decisions.

- Chapters 2-6 each describe a specific sequential decision problem to illustrate the modeling framework using a teach-by-example style. These applications were chosen to bring out each of the four classes of policies.

- Chapter 7 returns to the universal modeling framework in more detail. A much more careful discussion is given of the four classes of policies as well as different types of state variables, using the examples in Chapters 2-6 to provide context.

- Chapters 8-14 provide additional examples, using more complex settings to illustrate more advanced modeling concepts, covering both uncertainty modeling (in particular the modeling of electricity prices in Chapter 8) and a richer set of policies.

The application chapters (2-6 and 8-14) all follow the same outline. They can be covered in any order, keeping in mind that the applications in Chapters 2-6 are simpler and were chosen to illustrate each of the four classes of policies. Readers interested in specific modeling topics (such as state variables, modeling uncertainty, or seeing different examples of policies) may skim chapters, jumping directly to the topics that interest them.

Each chapter closes with a series of exercises divided into three categories:

- Review questions - These are simple questions that can be used to reinforce a basic understanding from reading the chapter.
- Problem solving questions - These introduce modeling challenges that require problem solving skills.
- Programming questions - Most chapters have programming exercises that draw on a set of Python modules which can be downloaded from the github directory.[1] Some of these questions require making programming modifications to the Python code.

1.2 The modeling process

Modeling is an art, but it is art guided by a mathematical framework that ensures that we get a well-defined problem that we can put on the computer and solve. This can be viewed as building a bridge from a messy, poorly defined real-world problem to something with the clarity a computer can understand, even if your end goal is not to put it on the computer.

In this section, we are going to provide a very compact version of our modeling framework. In Section 1.3, we are going to illustrate the framework, initially using a very simple inventory problem (in Section 1.3.1) but then introducing some modest extensions (in Section 1.3.2). After presenting these examples, we are going to return in Section 1.4 to a more detailed summary of our modeling framework.

1.2.1 A compact presentation of a dynamic model

We begin by observing that we can model any sequential decision problem using the sequence

$$(S_0, x_0, W_1, S_1, x_1, W_2, \ldots, S_t, x_t, W_{t+1}, \ldots, S_T),$$

where:

[1] https://tinyurl.com/sdagithub

- S_t is the *state variables* that captures what we know at time t. It is best to think of S_t as the state of information or, more generally, the state of knowledge, at time t. State variables can capture the vaccines in inventory, the location of a car, weather forecasts, and beliefs about how the market will respond to price.

- x_t represents *decision variables* which capture the elements that we control, such as whether to sell a house, the path through a network, the choice of drug for a treatment, the price at which to sell a product, or the assignment which trucks should move different loads of freight.

- W_{t+1} is the information that arrives after we make the decision x_t, which might be the final selling price of a house, the travel times through a network, how a patient responds to a drug, the sales of a product at a price, and the loads of freight called in after we make initial assignments. We view the information in W_{t+1} as coming from outside of our system, which means it is outside of our control. For this reason, we refer to it as *exogenous information.*

The decision x_t is determined by some method that we refer to as a *policy*, which we denote $X^\pi(S_t)$. We assume we have a *transition function* that takes as input the state S_t, decision x_t, and the exogenous information W_{t+1} and gives us the updated state S_{t+1}.

We incur a contribution (or cost) $C(S_t, x_t)$ when we make the decision $x_t = X^\pi(S_t)$ given the information in state S_t. Our goal is to find the policy that maximizes some objective that depends on the contributions $C(S_t, x_t)$ where $x_t = X^\pi(S_t)$.

There are many applications where it is more natural to index decisions using a counter n. In this case we would write

$$(S^0, x^0, W^1, S^1, x^1, W^2, \ldots, S^n, x^n, W^{n+1}, \ldots, S^N).$$

Here, n might be the nth experiment, the nth arrival of a customer, or the nth iteration of an algorithm.

Finally, there are settings where we use both, as in $(S_t^n, x_t^n, W_{t+1}^n)$ to capture, for example, decisions in the nth week at hour t.

This is a very compact description of a sequential decision problem. We next describe a set of steps to follow in the modeling process.

1.2.2 A six-step modeling process

It is possible to divide the entire modeling process into six steps (for our purposes):

Step 1. The narrative - This will be a plain English description of the problem. The narrative will not provide all the information needed to create a mathematical model; rather, it is a first step that should give the modeler the big picture without getting lost in notation.

Step 2. Identifying the core elements of the problem, with special emphasis on three dimensions of any sequential decision problem. These elements are described without using mathematics:

- What metrics are we trying to impact? Individual fields (such as supply chain management, health, energy, finance) will each be characterized by a number of metrics that might be described using terms such as costs, revenues, profits, rewards, gains, losses, performance and risk.
- What decisions are being made? Identifying decisions is quite easy for many problems such as computer games, but if we address a complex problem such as responding to a public health crisis, reducing the carbon footprint, or managing a supply chain, then identifying all the decisions can be quite challenging.
- What are the different sources of uncertainty? What are we uncertain about before we start? What information arrives exogenously over time (that is, arrives after we make a decision)? Table 1.2 illustrates different sources of uncertainty for a model of the distribution of COVID-19 vaccines (see Powell, 2022, Chapter 10, for a presentation of 12 classes of uncertainty).

Table 1.2: Illustration of different types of uncertainty arising in the vaccination response to the COVID-19 pandemic (adapted from Powell, 2022).

Type of uncertainty	Description
Observational errors	Observing people with symptoms Errors classifying people with symptoms as having COVID-19
Exogenous uncertainty	Reports of new cases, deaths Availability of ICUs Actual production of vaccines
Prognostic uncertainty	Hospital admissions Future performance of vaccines Population response to vaccines
Inferential uncertainty	Estimates of infection rates Estimates of effectiveness of vaccines
Experimental uncertainty	How a drug will perform in a clinical trial Number being vaccinated
Model uncertainty	Uncertainty in the transmission model used for forecasting Geographical spread of infections
Transitional uncertainty	Additions/withdrawals to/from vaccine inventories
Control uncertainty	Which population groups were vaccinated Vaccines allocations
Implementation uncertainty	Deviations when vaccines are not given to the correct people

Step 3. The mathematical model - Here we build off the first three elements from Step 2, but now we have to create a mathematical model that consists of five dimensions that apply to every sequential decision problem:

- State variables S_t - The state variable captures everything you need to know at time t to make a decision at time t, compute costs and constraints, and if necessary, simulate your way to time $t+1$. State variables can include information about physical resources (inventories or the location of a vehicle which enter the problem through constraints), other information (such as costs or prices which enter the objective function), and beliefs about quantities and parameters we do not know perfectly (such as forecasts or estimates of how a patient would respond to a drug).
- Decision variables x_t - These describe how we are going to design or control our system. Decisions have to satisfy constraints that we write as $x_t \in \mathcal{X}$ where $\mathcal{X}$ could be a set of discrete choices, or a set of linear equations. Decisions will be determined by *policies* that are functions (or rules) that we designate by $X^\pi(S_t)$ that determine x_t given what is in the state variable. Policies can be very simple (buy-low, sell-high) or quite complex. The index π carries information about the type of function that is used to make decisions (we return to this in considerable detail later in this chapter, and throughout the book). We introduce the policy $X^\pi(S_t)$, but defer until later designing the policy.
- Exogenous information W_{t+1} - This is new information that arrives after we make decision x_t (but before we decide x_{t+1}) such as how much we sell after setting a price, or the time to complete the path we chose. When we make a decision at time t, the information in W_{t+1} is unknown, so we treat it as a random variable when we are choosing x_t.
- The transition function $S^M(S_t, x_t, W_{t+1})$ - These are the equations that describe how the state variables evolve over time. For many real problems, transition functions capture all the dynamics of the problem, and can be quite complex. In some cases, we do not even know the equations, and have to depend on just the state variables that we can observe.

We write the evolution of the state variables S_t using our transition function as

$$S_{t+1} = S^M(S_t, x_t, W_{t+1}),$$

where $S^M(\cdot)$ is known as the state (or system) transition model (hence the M in the superscript). The transition function describes how every element of the state variable changes given the decisions x_t and exogenous information W_{t+1}. In complex problems, the transition function may require thousands of lines of code to implement.

- The objective function - This captures the performance metrics we use to evaluate our performance, and provides the basis for searching over policies. We let

  $C(S_t, x_t)$ = the contribution (if maximizing) or cost (if minimizing) of decision x_t which may depend on the information in S_t.

  Our objective is to find the best policy $X^\pi(S_t)$ to optimize some metric such as:

  – Maximize the expected sum of contributions over some horizon.
  – Maximize the expected performance of a final design that we learned over a number of experiments or observations.
  – Minimize the risk associated with a final design.

  Our most common way of writing the objective function is

$$\max_\pi \mathbb{E}\left\{\sum_{t=0}^{T} C(S_t, X^\pi(S_t))|S_0\right\}, \tag{1.4}$$

  where "$\mathbb{E}$" is called the *expectation operator* which means it is taking an average over anything random, which might include uncertain information in the initial state S_0, as well as the exogenous information process $W_1, \ldots, W_T$. It is standard to write the expectation operator, but we can never actually compute it. Later we show how to approximate it by running a series of simulations and taking an average, or by observing a process in the field.

The transition from the real problem (guided by the narrative) to the elements of the mathematical model is perhaps the most difficult step, as it often involves soliciting information from a non-technical source.

We note that we have presented the entire model without specifying how we make decisions, which is represented by the policy $X^\pi(S_t)$. We call this "*model first, then solve*" and it represents a major departure from the vast literature that deals with sequential decision problems. It is hard to communicate how important it is to approach sequential decision problems in this way.

Step 4. The uncertainty model - This is how we model the different types of uncertainty. There are two ways of introducing uncertainty into our model: 1) Through the initial state S_0 which might specify a probability distribution for uncertain parameters such as how a patient might respond to a drug or how the market might respond to price, and 2) through the exogenous information process $W_1, \ldots, W_T$.

We have three ways of modeling the exogenous information process:

- Create a mathematical model of $W_1, W_2, \ldots, W_T$.
- Use observations from history, such as past prices, sales, or weather events.
- Run the system in the field, observing W_t as they happen.

Step 5. Designing policies - Policies are functions, so we have to search for the best function. We do this by identifying two core strategies for designing policies:

- Search over a family of functions to find the one that works best, on average, over time.

- Create a policy by estimating the immediate cost or contribution of a decision x_t, plus an approximation of future costs or contributions, and then finding the choice x_t that optimizes the sum of current and future costs or contributions. Google maps decides whether to turn left or right by optimizing over the time required to transverse the next link in the network plus the remaining time to get to the destination. An inventory decision might optimize over the cost of an order plus the estimated value of holding a certain amount of inventory moving forward.

Step 6. Evaluating policies - Finding the best policy means evaluating policies to determine which is best. Most often we assume policies can be evaluated in a simulator, but simulators can be complex and difficult to build, and are still subject to modeling approximations. For this reason, the vast majority of practical problems encountered in practice tend to involve testing in the field, which introduces its own set of issues.

The only way to become comfortable with a mathematical model is to see it illustrated using a familiar example. We start with a universal problem that we all encounter in everyday life: managing inventories.

1.3 Some inventory problems

We are going to illustrate our modeling framework using two variations of a classic inventory problem, which is widely used as an application for illustrating certain methods for solving sequential decision problems. We start with a simple inventory example that gets across the core elements of our modeling framework, but allows us to ignore many of the complexities that we will be exploring in the remainder of the book.

Then, we are going to transition to a *slightly* more complicated inventory problem that will allow us to illustrate some modeling principles. Throughout the book, we are also going to use the idea of starting with a basic version of a problem, and then introduce extensions that hint at the types of complications that can arise in real applications.

1.3.1 A simple inventory problem

One of the most familiar sequential decision problems that we all experience each time we visit a store is an inventory problem. We are going to use a simple version of this problem to illustrate the six steps of our modeling process that we introduced above:

Step 1. Narrative - A pizza restaurant has to decide how many pounds of sausage to order from its food distributor. The restaurant has to make the decision at the end of day t, communicate the order which then arrives the following morning to meet tomorrow's orders. If there is sausage left over, it can be held to the following day. The cost of the sausage, and the price that it will be sold for the next day, is known in advance, but the demand is not.

Step 2. The core elements of the problem are:

- Metrics - We want to maximize profits given by the sales of sausage minus the cost of purchasing the sausage.
- Decisions - We have to decide how much to order at the end of one day, arriving at the beginning of the next.
- Sources of uncertainty - The only source of uncertainty in this simple model is the demand for sausage the next day.

Step 3. - The mathematical model - This consists of five elements.

1) The state variable S_t - We distinguish between the initial state variable S_0, and the dynamic state variable S_t for $t > 0$. The initial state variable S_0 consists of fixed parameters and initial values of variables that change over time, giving us

$$S_0 = (p, c, \bar{D}, \bar{\sigma}^D),$$

where p is the price we are paid for sausage, c is the purchase cost, and $(\bar{D}, \bar{\sigma}^D)$ is the mean and standard deviation of the demand which we assume is known from history. In more complex problems, there can be vast numbers of static parameters (such as the travel times over every link in a network) that are important, but which are not changing over time. We exclude these from the dynamic state variable which needs to be as compact as possible.

The dynamic state variable S_t is our inventory which we are going to call R_t^{inv}. For now, this is the only element of the dynamic state variable, so

$$S_t = R_t^{inv}.$$

Later we are going to introduce additional elements to our state variable.

2) The decision variable x_t is how much we order at time t, which we assume (for now) arrives right away. We make our decisions with a policy $X^\pi(S_t)$ which we design later.

3) The exogenous information is the random demand for our product which we are going to denote $\hat{D}_{t+1}$, so $W_{t+1} = \hat{D}_{t+1}$.

4) Our transition function captures how the inventory R_t evolves over time, which is given by

$$R_{t+1}^{inv} = \max\{0, R_t^{inv} + x_t - \hat{D}_{t+1}\}. \quad (1.5)$$

5) Our objective function. For our inventory problem, it is most natural to compute the contribution including the purchase cost of product x_t and the revenue from satisfying the demand $\hat{D}_{t+1}$, which means that our single-period contribution function would be written

$$C(S_t, x_t, \hat{D}_{t+1}) = -cx_t + p\min\{R_t + x_t, \hat{D}_{t+1}\},$$

where $x_t = X^\pi(S_t)$. Given a sequence of demands $\hat{D}_1, \ldots, \hat{D}_T$, the value of a policy $\hat{F}^\pi$ would be

$$\hat{F}^\pi = \sum_{t=0}^{T} C(S_t, X^\pi(S_t), \hat{D}_{t+1}).$$

Our profits $\hat{F}^\pi$ are random because it depends on a particular sequence of random demands $\hat{D}_1, \ldots, \hat{D}_T$. Finally we average over these random demands by taking the expectation:

$$F^\pi = \mathbb{E}\left\{\sum_{t=0}^{T} C(S_t, X^\pi(S_t), \hat{D}_{t+1})|S_0\right\}. \tag{1.6}$$

Here, the conditioning on the initial state S_0 can be read as saying "take the expectation given what we know initially." Conditioning on S_0 is implicit any time we take an expectation, and as a result many authors leave it out. However, we are going to include the conditioning on S_0 to make it clear that if our initial inputs (including beliefs) change, then this may have an effect on how a policy performs.

Step 4. The uncertainty model - The simplest approach to modeling uncertainty is to just use historical data. The problem we might encounter is that if we run out of sausage, we might not observe the full demand for sausage that day. If we are able to capture this lost demand, then this is a reasonable approach.

An alternative is to build a mathematical model. We might assume that our demand is normally distributed with some mean $\bar{D}$ and standard deviation $\bar{\sigma}^D$. If we assume that both of these are known, we can write our demand as

$$\hat{D}_{t+1} \sim N(\bar{D}, (\bar{\sigma}^D)^2),$$

and take advantage of packages that can sample from the normal distribution (for example, in Excel this is called $\texttt{Norm.inv}(Rand(), \bar{D}, \bar{\sigma})$) to generate a random observation with mean $\bar{D}$ and standard deviation $\bar{\sigma}$.

Using this model, we can create a set of demands $(\hat{D}_1, \hat{D}_2, \ldots, \hat{D}_T)$. Then, we can repeat this N times to create N sequences of T demands, giving us the sequence $(\hat{D}_1^n, \hat{D}_2^n, \ldots, \hat{D}_T^n)$ for $n = 1, \ldots, N$ that we need to estimate the value of the policy (we use this below in Step 6).

Step 5. Designing policies - Next we have to design a method for determining our orders. A commonly used strategy for inventory problems is known as an "order-up-to" policy that looks like

$$X^\pi(S_t|\theta) = \begin{cases} \theta^{max} - R_t & \text{if } R_t < \theta^{min}, \\ 0 & \text{otherwise,} \end{cases} \tag{1.7}$$

where $\theta = (\theta^{min}, \theta^{max})$ is a set of parameters that need to be tuned. It is called "order-up-to" since we place an order to bring the inventory "up to" the upper limit θ^{max}.

Step 6. Evaluating policies - There are a variety of strategies we might use. In practice, we cannot compute the expectation in the objective function in Equation (1.6), so we take a series of samples of demands. Let $\hat{D}_1^n, \ldots, \hat{D}_T^n$ be one sample of demands over $t = 1, \ldots, T$, and assume we can generate N of these. Now we can estimate our expected profits from policy $X^\pi(S_t)$ by averaging over the samples for $n = 1, \ldots, N$, which is computed using

$$\overline{F}^\pi(\theta) = \frac{1}{N} \sum_{n=1}^{N} \sum_{t=0}^{T} C(S_t, X^\pi(S_t|\theta), \hat{D}_{t+1}^n).$$

In plain English, we are simulating the policy $X^\pi(S_t|\theta)$ N times using the simulated (or observed from history) samples of demands $\hat{D}_1^n, \ldots, \hat{D}_T^n$, and then averaging the performance to get $\overline{F}^\pi(\theta)$. We then face the problem of finding the best value of θ. A simple strategy would be to generate K possible values $\theta_1, \ldots, \theta_K$, simulating each one to find $\overline{F}^\pi(\theta_k)$ for each k, and then pick the value of θ_k that works the best. This is not an optimal strategy, but it provides a simple, practical starting point.

1.3.2 A slightly more complicated inventory problem

The simple inventory problem above is a classic setting for demonstrating a particular method for solving sequential decision problems known as dynamic programming, which depends on having a simple state variable that is a) discrete and b) does not have too many possible values. In our slightly more complicated inventory problem, we are going to illustrate three different flavors of state variables which would represent a serious complication for one popular method for solving sequential decision problems, but has no effect on the policy that we have chosen.

Step 1. Narrative - We again have our pizza restaurant that has to order sausage, but we are going to allow the price we pay for sausage to vary from day to day, where we assume the price on one day is independent of the price on the previous day. Then, we are also going to assume that while the demand for sausage tomorrow is random, we are going to be given a forecast of tomorrow's demand that, while not perfect, is better than not having a forecast. Otherwise, everything about our more complicated problem is the same as it was before.

Step 2. Core elements - These are:

- Metrics - We want to maximize profits given by the sales of sausage minus the cost of purchasing the sausage, where the cost varies from day to day.
- Decisions - As with our simpler inventory problem, we have to decide how much to order at the end of one day, arriving at the beginning of the next.
- Sources of uncertainty - There are now three sources of uncertainty: the difference between the actual demand and the forecast, the evolution of the forecasts from one day to the next, and the price we pay for the sausage.

Step 3. - Mathematical model - We still have the same five elements, but now the problem is a bit richer:

1) To construct the state variable, we need to list the information (specifically, information that evolves over time) that is needed in three different parts of the model:

1) The objective function.
2) The policy for making decisions (which includes the constraints).
3) The transition function.

Of course, we have not yet introduced any of these functions, so you have to read forward, and verify that our state variable contains all the information needed to compute each of these functions. Think of this as a dictionary of the information we will need.

We start with the initial state S_0 which consists of constant parameters, and initial values of quantities and parameters that change over time. These are:

- Price - We assume that we sell our sausage at a fixed price p.
- Initial inventory - We start with an initial inventory R_0.
- Initial purchase cost - c_0.
- Initial forecast - We assume that our first forecast $f^D_{0,1}$ is given, where $f^D_{0,1}$ is the forecast known at time 0 for the demand at time 1.
- Initial estimate of the standard deviation of the demand - $\bar{\sigma}^D_0$.
- Initial estimate of the standard deviation of the forecast - $\bar{\sigma}^f_0$.

This means our initial state variable is

$$S_0 = (p, R_0, c_0, f^D_{0,1}, \bar{\sigma}^D_0, \bar{\sigma}^f_0).$$

We then have the information that evolves over time which makes up our dynamic state variable S_t:

- Current inventory R^{inv}_t - The inventory for the beginning of time interval $(t, t+1)$.
- Purchase cost c_t - This is the cost of sausage purchased at time t which is given to us at time t.
- Demand forecast $f^D_{t,t+1}$ - This is the forecast of $\hat{D}_{t+1}$ given what we know at time t.
- Current estimate of the standard deviation of the demand - $\bar{\sigma}^D_t$.
- Current estimate of the standard deviation of the forecast - $\bar{\sigma}^f_t$.

Our dynamic state variable is then given by

$$S_t = (R^{inv}_t, c_t, f^D_{t,t+1}, \bar{\sigma}^D_t, \bar{\sigma}^f_t).$$

2) The decision variable x_t is how much we order at time t, which we assume (for now) arrives right away. We make our decisions with a policy $X^\pi(S_t)$ which we design later.

3) The exogenous information now consists of:

- Purchase costs $\hat{c}_{t+1}$ - This is the purchase cost of sausage on day $t+1$ which is specified exogenously.
- Forecasts - Each time period we are given a new forecast. Let ε^f_{t+1} be the change in the forecast between time t and $t+1$.
- Demands - Finally, we assume that the actual demand is a random deviation from the forecast, which we could write

$$\hat{D}_{t+1} = f^D_{t,t+1} + \varepsilon^D_{t+1}.$$

Our complete set of exogenous information variables can now be written

$$W_{t+1} = (\hat{c}_{t+1}, \varepsilon^f_{t+1}, \varepsilon^D_{t+1}).$$

4) Transition function - This specifies how each of the (dynamic) state variables S_t evolve over time. We update our inventory using:

$$R^{inv}_{t+1} = \max\{0, R^{inv}_t + x_t - \hat{D}_{t+1}\}. \quad (1.8)$$

The demand is the forecasted demand plus the deviation ε^D_{t+1} from the forecast, giving us the equation

$$\hat{D}_{t+1} = f^D_{t,t+1} + \varepsilon^D_{t+1}. \quad (1.9)$$

We assume that our forecast is updated using

$$f^D_{t+1,t+2} = f^D_{t,t+1} + \varepsilon^f_{t+1}. \quad (1.10)$$

Next, we are going to adaptively estimate the variance in the demand and the demand forecast

$$(\bar{\sigma}^D_{t+1})^2 = (1-\alpha)(\bar{\sigma}^D_t)^2 + \alpha(f^D_{t,t+1} - \hat{D}_{t+1})^2, \quad (1.11)$$
$$(\bar{\sigma}^f_{t+1})^2 = (1-\alpha)(\bar{\sigma}^f_t)^2 + \alpha(f^D_{t,t+1} - f^D_{t+1,t+2})^2, \quad (1.12)$$

where $0 < \alpha < 1$ is a smoothing factor.

Finally, we update the cost c_{t+1} with the "observed cost" $\hat{c}_{t+1}$ which we write simply as

$$c_{t+1} = \hat{c}_{t+1}. \tag{1.13}$$

Equation (1.13) is an example of a state variable that we observe rather than compute, as we did with the inventory R_t^{inv} in (1.8). Equation (1.8) is sometimes referred to as "model based" while Equation (1.13) is called "model free."

Our transition function

$$S_{t+1} = S^M(S_t, x_t, W_{t+1})$$

consists of the Equations (1.8) - (1.13).

5) Finally, our single period contribution function would now be written

$$C(S_t, x_t, \hat{D}_{t+1}) = -c_t x_t + p \min\{R_t + x_t, \hat{D}_{t+1}\},$$

where the only difference with the simpler inventory problem is that the cost c is now time-dependent c_t. We break from our convention of writing the contribution as $C(S_t, x_t)$ and allow it to include revenues from the demands $\hat{D}_{t+1}$.

We now state our objective function formally as

$$\max_\pi \mathbb{E}\left\{\sum_{t=0}^{T} C(S_t, X^\pi(S_t), \hat{D}_{t+1})|S_0\right\}. \tag{1.14}$$

The optimization $\max_\pi$ means that we are searching over all possible policies, which literally means searching over all the different functions we might use to make a decision. Recall that above we stated that the index π carries information about the type of function $f \in \mathcal{F}$, and any tunable parameters $\theta \in \Theta^f$. In practice, the search over the types of functions $f \in \mathcal{F}$ tends to be ad hoc (a knowledgeable analyst chooses functions that make sense for a problem), whereas a computer algorithm performs the search for the best value of $\theta \in \Theta^f$.

Section 1.6 goes into the challenge of designing policies in more detail. In fact, a substantial portion of this book is dedicated to illustrating different types of policies in the context of different applications.

Step 4. The uncertainty model - We are going to assume that the exogenous changes ε^D_{t+1} and ε^f_{t+1} are described by normal distributions with mean 0 and variances $(\bar{\sigma}^D_t)^2$ and $(\bar{\sigma}^f_t)^2$, which we express by writing

$$\begin{aligned}\varepsilon^D_t &\sim N(0, (\bar{\sigma}^D_t)^2),\\ \varepsilon^f_t &\sim N(0, (\bar{\sigma}^f_t)^2).\end{aligned}$$

Uncertainty models can become quite complex, but this will serve as an illustration.

Step 5. Designing policies - Next we have to design a method for determining our orders. Instead of the order-up-to policy of our simpler model, we are going to suggest the idea of ordering enough to meet the expected demand for tomorrow, with an adjustment. We could write this as

$$X^\pi(S_t|\theta) = \max\{0, f^D_{t,t+1} - R_t\} + \theta. \tag{1.15}$$

If we had a perfect forecast, then all we have to order would be $f^D_{t,t+1}$ (our forecast of $\hat{D}_{t+1}$) minus the on-hand inventory. However, because of uncertainty we are going to add an adjustment (presumably $\theta > 0$, but this depends on the sales price relative to the purchase cost).

Step 6. Evaluating policies - This time we have to generate samples of all the random variables in the sequence $W_1, W_2, \ldots, W_T$. Again we might generate N samples of the entire sequence so we can estimate the performance of a policy using

$$\overline{F}^\pi(\theta) = \frac{1}{N}\sum_{n=1}^{N}\sum_{t=0}^{T} C(S_t, X^\pi(S_t|\theta), \hat{D}^n_{t+1}).$$

We again face the problem of finding the best value of θ, but we return to that challenge later.

1.4 The mathematical modeling framework

Section 1.2 provided a streamlined summary of the modeling process, but did not provide some of the details needed to tackle more complex problems than the simple illustrations in Section 1.3. Below we provide a more detailed summary of the elements of a sequential decision problem.

State variables - The state S_t of the system at time t has all the information that is necessary and sufficient to model our system from time t onward. This means it has to capture the information needed to compute costs/contributions, constraints on decisions, and the evolution of this information over time in the transition function. In fact, it is often best to write out the objective function (with costs/contributions), the constraints for the decisions, and the transition function, and then return to define the state variables.

There will be many settings where it makes more sense to use a counter n rather than time. In this case, we let S^n be the state after n observations (these may be experiments, customer arrivals, iterations of an algorithm). We will use time t as our default index.

We need to distinguish between the initial state S_0 and subsequent states S_t for $t > 0$:

S_0 - The initial state S_0 captures i) deterministic parameters that never change, ii) initial values of quantities or parameters that do change (possibly due to decisions), and iii) beliefs about quantities or parameters that we do not know perfectly (this might be the parameters of a probability distribution) such as how we respond to a vaccine or how the market will respond to price.

S_t - This is all the information we need at time t from history to model the system from time t onward.

There are three types of information in S_t:

- The physical state, R_t, which in most (but not all) applications is the state variable that is being controlled. R_t may be a scalar, or a vector with element R_{ti} where i could be a type of resource (e.g. a blood type) or the amount of inventory at location i. In our inventory problems, the physical state variables would be the inventory R_t and the demands $\hat{D}_t$.

- Other information, I_t, which is any information that is known deterministically not included in R_t, such as prices or the current temperature. The information state often evolves exogenously, but may depend on decisions.
- The belief state, B_t, which contains information about a probability distribution describing one or more unknown parameters. Thus, B_t could capture the mean and variance of a normal distribution (as with our demand forecast above). Alternatively, it could be a vector of probabilities that evolve over time.

We note that the distinction of "physical state" R_t versus "other information" I_t is somewhat imprecise, but there are many problems that involve the management of physical (and financial) resources where the state of the physical resources (inventories, location of a vehicle, location and speed of a robot) is a natural "state variable" (in fact, a common error is to assume that the state of physical resources is the only state variable). By contrast, the "other information" I_t is literally any other information that does not seem to belong in R_t.

For example, R_t might be the amount of money in a cash account, while I_t might be the current state of the stock and bond markets. If we are traveling over a dynamic network, R_t might be our location on the network, while I_t could be what we know about the travel times over each link. If we plan a path and then wish to penalize deviations from the plan, then the plan would be included in the state variable through I_t.

State variables typically are not obvious. They emerge during the modeling process, rather than something you can just immediately write out. Just because we write it first does not mean that you will always be able to list all the elements of the state variable right away. But in the end, this is where you store all the information you need to model your system from time t onward.

Decision variables - Different communities use different notations for decision, such as a_t for a (typically discrete) action or u_t for a (typically continuous) control in engineering. We use x_t as our default since it is widely used by the math programming community.

Decisions may be binary (e.g. for modeling whether to sell an asset or not, or for A/B testing of different web designs), discrete (e.g. choice of drug, which product to advertise), continuous (prices, temperatures, concentrations), vectors (discrete or continuous such as allocations of blood supplies among hospitals), and categorical (e.g. what features to highlight in a product advertisement). We note that entire fields of research are sometimes distinguished by the nature of the decision variable.

We assume that decisions are made with a policy, which we might denote $X^\pi(S_t)$ if we use x_t as our decision. We assume that a decision $x_t = X^\pi(S_t)$ is feasible at time t, which means $x_t \in \mathcal{X}_t$ for some set (or region) $\mathcal{X}_t$, which may depend on S_t.

We let "π" carry the information about the type of function $f \in \mathcal{F}$ (for example, a linear model with specific explanatory variables, or a particular nonlinear model), and any tunable parameters $\theta \in \Theta^f$.

Exogenous information - We let W_{t+1} be any new information that first becomes known at time $t+1$ (that is, between t and $t+1$), where the source of the information is from outside of our system (which is why it is "exogenous"). When modeling specific variables, we use "hats" to indicate exogenous information. Thus, $\hat{D}_{t+1}$ could be the demand that arises between t and $t+1$, or we could let $\hat{p}_{t+1}$ be the change in the price between t and $t+1$.

The exogenous information process may be stationary or nonstationary, purely exogenous or state (and possibly action) dependent (if we decide to sell a lot of stock, it could push prices down).

We let ω represent a sample path $W_1, \ldots, W_T$, which represents a sequence of outcomes of each W_t. Often, we will create a set Ω of discrete samples, where each sample represents a particular sequence of the outcomes of our W_t process which we could write as $W_1(\omega), \ldots, W_T(\omega)$. If we have 20 sample paths, we can think of ω as consisting of a number between 1 and 20, which allows us to look up the sample path.

Transition function - We denote the transition function by

$$S_{t+1} = S^M(S_t, x_t, W_{t+1}), \tag{1.16}$$

where $S^M(\cdot)$ is also known by names such as state transition model, system model, plant model, plant equation, state equation, and transfer function.

Equation (1.16) is the classical form of a transition function which gives the equations from the state S_t to the state S_{t+1}. Equation (1.5) was the only transition equation for our simple inventory example, while Equations (1.8) - (1.13) made up the transition function for our more complicated example.

The transition function might capture any of the following types of updates:

- Changes in physical resources such as adding inventory, moving people, or modifying equipment.
- Updates to information such as changes in prices and weather.
- Updates to our beliefs about uncertain quantities or parameters.

The transition function may be a known set of equations, or unknown, such as when we describe human behavior or the evolution of $CO2$ in the atmosphere. When the equations are unknown the problem is often described as "model free" or "data driven" which means we can only observe changes in a variable, rather than using a physical model. Equation (1.13), where we "observe" the cost $c_{t+1} = \hat{c}_{t+1}$, with no idea how we evolved from c_t, is an example of a model free transition.

Transition functions may be linear, continuous nonlinear or step functions. When the state S_t includes a belief state B_t, then the transition function has to include the updating equations (we illustrate this later in the book).

Given a policy $X^\pi(S_t)$, an exogenous process W_{t+1}, and a transition function, we can write our sequence of states, decisions, and information as

$$(S_0, x_0, W_1, S_1, x_1, W_2, \ldots, x_{T-1}, W_T, S_T).$$

Objective functions - There are a number of ways to write objective functions. One of the most common, which we will use as a default, maximizes total expected contributions over some horizon $t = 0, \ldots T$

$$\max_\pi \mathbb{E}\left\{\sum_{t=0}^{T} C_t(S_t, X_t^\pi(S_t))|S_0\right\}, \tag{1.17}$$

where

$$S_{t+1} = S^M(S_t, X_t^\pi(S_t), W_{t+1}). \tag{1.18}$$

The model is fully specified when we also have a model of the initial state S_0, and a model of the exogenous process $W_1, W_2, \ldots$. We write all the exogenous information as

$$(S_0, W_1, W_2, \ldots, W_T). \tag{1.19}$$

Equations (1.17), (1.18) and (1.19) constitute a model of a sequential decision problem.

Equation (1.17) uses an expectation $\mathbb{E}$ which means to take an average over all the possible outcomes of $W_1, \ldots, W_T$. This is virtually never possible to do computationally. Instead, let ω represent a single outcome of the sequence $W_1, \ldots, W_T$ which we might write $W_1(\omega), \ldots, W_T(\omega)$. Assume that we can create N possible outcomes of this sequence, and let ω^n represent how we index the nth sequence.

If we are following a sample path ω, we would then rewrite our transition function in (1.18) using

$$S_{t+1}(\omega) = S^M(S_t(\omega), X_t^\pi(S_t(\omega)), W_{t+1}(\omega)). \tag{1.20}$$

We index every variable in Equation (1.20) by ω to indicate that we are following a single sample path of values of W_t.

We can now replace our expectation-based objective with an average which we can write

$$\max_\pi \frac{1}{N} \sum_{n=1}^{N} \sum_{t=0}^{T} C_t(S_t(\omega^n), X_t^\pi(S_t(\omega^n))). \tag{1.21}$$

Any time we write an objective using an expectation as in (1.17), remember that what we would really do is to use an average as we do in (1.21).

Note that the objective function depends on the initial state S_0. As described earlier, the initial state includes any deterministic parameters, initial values of dynamic parameters (such as inventory or the initial location of a drone), or the initial values of a probability distribution describing an uncertain quantity (such as the health of a patient). Whatever we compute using the objective function in (1.17) depends, at least implicitly, on the information in S_0.

At this point, we have a model, but we are not done. We still have two major steps remaining:

- Modeling uncertainty - Typically we describe the elements of the initial state S_0, and the exogenous information process $W_1, \ldots,$ W_T, but we have to find some way to generate samples from the underlying probability distributions.

- Designing policies - This will be a focal point of this book. When we described our inventory problems, we suggested very simple policies just to provide an idea of a policy. However, our standard method is to "model first, then solve" which means to specify the model using the function $X^\pi(S_t)$ as a placeholder. Then, once we have created our model (including the uncertainty model), we turn to the challenge of designing policies.

1.5 Modeling uncertainty

For many complex problems (supply chains, energy systems, and public health are just a few), identifying and modeling the different forms of uncertainty can be a rich and complex exercise. We are going to hint at the issues that arise, but we are not going to attempt a thorough discussion of this dimension.

Uncertainty is communicated to our model through two mechanisms: the initial state S_0, which is where we would model the parameters of probability distributions describing quantities and parameters that we do not know perfectly, and the exogenous information process $W_1, \ldots, W_T$.

1.5.1 The initial state variables S_0

The initial state variable may contain deterministic parameters or initial values of dynamically varying quantities and parameters. If this is all that is in the initial state, then it is not capturing any form of uncertainty.

There are many problems where we do not know some quantities or parameters, but can represent what we do know through the parameters of a probability distribution. Some examples are:

- The response of a patient to a new medication.
- How a market will respond to a change in price.
- How many salable heads of lettuce we have in inventory (an uncertain number may have wilted and are no longer salable).
- The time at which a box of inventory previously ordered from China will arrive.
- The amount of deposits to a mutual fund vary randomly around a mean λ, but we do not know what λ is.

These are a number of ways that we can initialize a model with uncertainty in some of the inputs.

An initial probabilistic belief may come from subjective judgment, or from previous observations or experiments.

1.5.2 The exogenous information process

The second way uncertainty enters our model is through the exogenous information process. The variable W_t contains information that is not known until time period t. This means we have to make a decision x_t at time t before we know the outcome of W_{t+1}.

Below is a list of examples of W_{t+1} that are revealed after a decision x_t is made:

- We choose a path, and then observe the travel time on the path.
- We choose a drug, and then observe how the patient responds.
- We choose a catalyst, and then observe the strength of the material that it produces.
- We choose a product to advertise in an online market, and then observe the sales.
- We select a web interface design, and then observe the number of clicks that it can generate.
- We allocate funds into an investment, and then observe the change in the price of the investment.

In each case, the information we observe after we make the decision affects the performance of the decision (and which decision would have been best).

We note that the information W_{t+1} helps to determine the state S_{t+1}, which in turn is used to make the following decision x_{t+1}. This means that the decision x_{t+1} is uncertain at time t when we are making our decision x_t.

1.5.3 Testing policies

To test the value of a policy, we need a way to generate samples of the exogenous information process. We let ω be a sample path, where $W_1(\omega), \ldots, W_T(\omega)$ represents a particular sample path. Table 1.3

illustrates 10 sample paths of prices that are indexed ω^1 to ω^{10}. If we choose ω^6, then

$$W_7(\omega^6) = 44.16.$$

Table 1.3: Illustration of a set of sample paths for prices all starting at \$45.00.

	$t=1$	$t=2$	$t=3$	$t=4$	$t=5$	$t=6$	$t=7$	$t=8$
ω^n	p_1	p_2	p_3	p_4	p_5	p_6	p_7	p_8
ω^1	45.00	45.53	47.07	47.56	47.80	48.43	46.93	46.57
ω^2	45.00	43.15	42.51	40.51	41.50	41.00	39.16	41.11
ω^3	45.00	45.16	45.37	44.30	45.35	47.23	47.35	46.30
ω^4	45.00	45.67	46.18	46.22	45.69	44.24	43.77	43.57
ω^5	45.00	46.32	46.14	46.53	44.84	45.17	44.92	46.09
ω^6	45.00	44.70	43.05	43.77	42.61	44.32	44.16	45.29
ω^7	45.00	43.67	43.14	44.78	43.12	42.36	41.60	40.83
ω^8	45.00	44.98	44.53	45.42	46.43	47.67	47.68	49.03
ω^9	45.00	44.57	45.99	47.38	45.51	46.27	46.02	45.09
ω^{10}	45.00	45.01	46.73	46.08	47.40	49.14	49.03	48.74

The question is: how do we create a sample of observations such as those depicted in Table 1.3? There are three typical strategies:

- Create samples from historical data. Since there is only one outcome at any point in time, we can create multiple sample paths by combining observations from different periods of time. We might pick prices from different years, or demands from different months, or observed travel times on different days.

- We can test an idea in the field, using observations as they actually occur. The advantage of this is that we are working with real data (history may not be the same as the future). The disadvantage is that it takes a day to observe a day of new data (and we may need much more than a single day of observations).

- Simulating from a mathematical model. This approach offers the advantage of being able to generate large samples to get statistically reliable estimates of the performance of a policy. These models can be quite sophisticated, but it is quite easy to create models (even sophisticated ones) that do not replicate the behavior of real data. The biggest challenge is capturing correlations, either over time or between samples (say the demands of different products, the prices of different stocks, or the wind speed in different locations).

We refer the reader to Powell (2022, Chapter 10) for a more in-depth discussion of uncertainty modeling.

1.6 Designing policies

Throughout the applications in the chapters that follow, we will always be trying to find a good policy, which might be the best in a class, or sometimes the best overall (the true optimal policy). Figure 1.2 shows the front covers of books representing roughly 15 distinct fields that all deal with sequential decisions under uncertainty. These books are largely distinguished by how they approach creating policies, along with differences in problem characteristics.

We are going to build on the idea that there are two fundamental strategies for creating policies, each of which can then be further divided into two classes, creating four classes of policies:

Figure 1.2: A sampling of major books representing different fields in stochastic optimization.

Policy search - This is where you search across methods (functions) for making decisions, simulating their performance (as we do in Equation (1.17)), to find the method that works best on average over time. This may involve searching over different classes of methods, as well as any tunable parameters for a given method. This idea opens up two classes of policies:

1) Policy function approximations (PFAs) - These are analytical functions of a state that directly specify an action. The order-up-to policy in Equation (1.7) is a good example, along with our policy of using an adjusted forecast in Equation (1.15).

2) Cost function approximations (CFAs) - These are policies that involve solving an optimization problem that is typically a simplification of the original problem, with parameters introduced to help make the policy work better over time. This is a particularly powerful idea that is widely used in industry. We have a number of illustrations of CFAs later in the book (starting with Chapter 4 to learn the best medication for diabetes).

Lookahead policies - We can build effective policies by optimizing across the contribution (or cost) of a decision, plus an approximation of the downstream contributions (or cost) resulting from the decision made now. Again, we can divide these into two more classes of policies:

3) Value function approximations (VFAs) - Imagine that we are traversing a network depicted in Figure 1.3 where we wish to find a path from node 1 to node 11. Now imagine that we are at node $S_t = i = 2$, where t counts how many links we have traversed. Let $V_{t+1}(S_{t+1})$ be the value (assuming we are maximizing) of the path from node S_{t+1} (such as nodes 4 or 5) to node 11 (don't worry about how we obtained $V_{t+1}(S_{t+1})$). Let a decision x_t be the link we traverse out of node $S_t = i$. The value of being at node S_t would be given

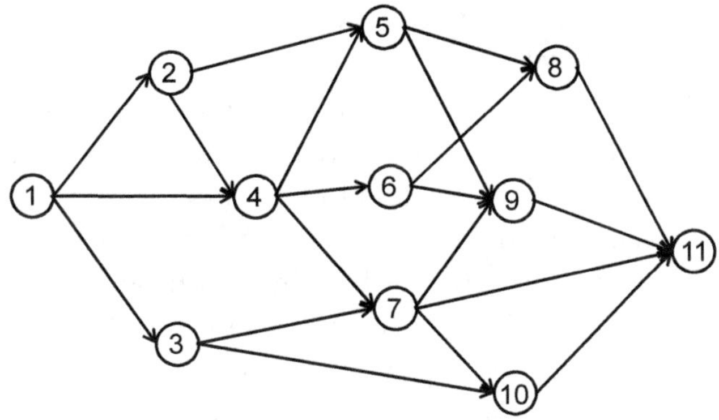

Figure 1.3: Simple deterministic graph for traversing from node 1 to node 11.

by

$$V_t(S_t) = \max_{x_t} \left(C(S_{t,t}\, x) + V_{t+1}(S_{t+1}) \right). \tag{1.22}$$

Equation (1.22) is known as *Bellman's equation*. When it is used to find the best path in a deterministic network such as the one we depicted in Figure 1.3, it is fairly easy to visualize. There are many problems where the transition from state S_t to S_{t+1} involves random information that is not known at time t. We saw a simple example of randomness in our first inventory problem in Section 1.3.1, and a more complicated example in Section 1.3.2.

For these more general problems, if we are in a state S_t, make a decision x_t and then observe new information W_{t+1} (that is not known at time t) it will take us to a new state S_{t+1} according to our transition function

$$S_{t+1} = S^M(S_t, x_t, W_{t+1}). \tag{1.23}$$

This means that at time t when we have to choose x_t, W_{t+1} is a random variable, which means that S_{t+1} is a random variable as well. In this case we have to insert an expectation in Bellman's equation and write Equation (1.22) as

$$V_t(S_t) = \max_{x_t} \left(C(S_{t,t}\, x) + \mathbb{E}\left\{ V_{t+1}(S_{t+1}) | S_t, x_t \right\} \right). \tag{1.24}$$

Here we have inserted the expectation $\mathbb{E}\{\cdot\}$ which literally means to average over all the random outcomes of W_{t+1}.

The stochastic version of Bellman's equation in (1.24) is extremely general. The state S_t does not just mean a node in a graph; it captures any (and all) information relevant to the problem. The difficulty is that we can no longer compute the value function $V_t(S_t)$, which in turn means we will not have access to $V_{t+1}(S_{t+1})$ that we assumed we knew in Equations (1.22) and (1.24).

The strategy the research community has used when trying to apply Bellman's equation is to draw on the field of machine learning to estimate a statistical approximation that we are going to call $\overline{V}_t(S_t)$. Assuming that we can come up with a reasonable approximation $\overline{V}_{t+1}(S_{t+1})$, we would write our policy (our method for making a decision) using

$$X^\pi(S_t) = \arg\max_{x_t \in \mathcal{X}_t} \left(C(S_t, x_t) + \mathbb{E}\{\overline{V}_{t+1}(S_{t+1})|S_t, x_t\} \right). \quad (1.25)$$

The notation $\arg\max_x f(x)$ means the value of x that maximizes the function $f(x)$.

This class of policy falls under headings such as approximate dynamic programming and, most often, reinforcement learning. While a powerful idea, it is not easy to apply and depends on our ability to create an accurate approximation $\overline{V}_{t+1}(S_{t+1})$.

There is a very rich literature on methods for approximating value functions, but it is no panacea. This book will illustrate this idea in a few places.

4) Direct lookahead approximations (DLAs) - There are many problems where we simply cannot develop effective policies using any of the first three classes, and when this happens, we have to turn to direct lookahead approximations. We will write this out in its full mathematical form later, but for now, we are going to describe DLAs as making a decision

now while optimizing over a (typically approximate) model that extends over some planning horizon.

A common DLA is to create an approximate model that is deterministic. This is what we are doing when we use a navigation system that finds the shortest path to the destination assuming that we know the travel time along each link of the network. As a general rule, solving an exact stochastic model of the future is almost always impossible, so we are going to investigate different strategies for approximating the problem.

We illustrated our modeling framework in Section 1.3 using two inventory problems, and suggested two simple policies (forms of PFAs) with Equations (1.7) and (1.15), but we did this just to have a concrete example of a policy. While PFAs are widely used in day-to-day decision making, these are specialized examples.

By contrast, we are going to claim that the four classes of policies we just outlined (PFAs, CFAs, VFAs and DLAs) are universal, in that these cover *any* method that we might use to solve *any* sequential decision problem. To be clear, these are meta-classes. That is, if we think a problem lends itself to one particular class, we are not done, since we still have to design the specific policy within the class. Just the same, we feel that these four classes provides a roadmap to guide the process for designing policies.

1.7 Next steps

The next five chapters of the book are going to apply our modeling framework to five different problems:

- Chapter 2 - An asset selling problem
- Chapter 3 - Adaptive market planning
- Chapter 4 - Learning the best diabetes medication
- Chapter 5 - Stochastic shortest path problems - Static
- Chapter 6 - Stochastic shortest path problems - Dynamic

Each of these chapters will follow the same outline as we used in Section 1.3 to describe the two inventory problems. This outline consists of:

- Narrative - A plain English description of the problem.
- Model - This model will follow our format of describing the five elements of a sequential decision problem: state variables, decision variables, exogenous information variables, the transition function and the objective function.
- Uncertainty model - Here we will provide a possible model of any uncertainties in the problem.
- Designing policies - We are going to suggest possible policies for making decisions. We have chosen our problems so that the five application settings will give us a tour through all four classes of policies. For now, we are going to let the reader try to recognize which of the four classes we are choosing.
- Extensions - Finally we may suggest one or more possible extensions of our basic problem that may require changing the policy.

We then return to the four classes of policies in Chapter 7 and discuss our general modeling framework, using the problems in Chapters 2 - 6 to illustrate different modeling ideas.

After this discussion, we return to our pattern of teach-by-example chapters, but using more complex problems. Our remaining chapters cover the following problems:

- Chapter 8 - Energy storage I
- Chapter 9 - Energy storage II
- Chapter 10 - Supply chain management I: The two-agent newsvendor
- Chapter 11 - Supply chain management II: The beer game
- Chapter 12 - Ad-click optimization
- Chapter 13 - Blood management problem
- Chapter 14 - Optimizing clinical trials

1.8 What did we learn?

- We presented a general model for any sequential decision problem.
- We illustrated the model by first using a classical inventory problem, where the state of the system is the amount in inventory.
- We then transitioned to a slightly more complicated inventory problem where the state variable includes the resource state R_t of inventory, an informational state variable in the form of price p_t, and finally a belief state about the upcoming demand $\hat{D}_{t+1}$ in the form of an estimated mean and variance.
- We illustrated two forms of a simple class of policy known as a policy function approximation (or PFA).
- We closed with a brief overview of four classes of policies. Illustrations of all four classes will be provided in Chapters 2-6, at which point we pause in Chapter 7 to discuss the policies in greater depth, setting the stage for the more complex problems in Chapters 8-14.

1.9 Exercises

Review questions

1.1 — What are the five elements of the mathematical model of a sequential decision problem?

1.2 — What is the difference between the variables in the initial state S_0 and those in the dynamic state S_t for $t > 0$?

1.3 — What is the difference between a decision and the exogenous information?

1.4 — What are the two major categories of policies, and how are they different?

1.5 — Compare the state variables for the simple inventory problem in Section 1.3.1 to the more complicated inventory problem in Section 1.3.2.

Problem solving questions

1.6 — Compare the policies for the two inventory problems in Section 1.3.1 in terms of how they would handle time-dependent behavior. For example, our pizza parlor may have much higher demands on weekends than on weekdays. Comment on the value of making the tunable parameter θ time-dependent (or day of week dependent) in terms of how it might improve the solution.

1.7 — Contrast how you might go about tuning the parameter θ for the inventory problems in Section 1.3.1:

a) In a simulator.

b) In the field.

Discuss the pros and cons of each approach.

2

An Asset Selling Problem

2.1 Narrative

We are holding a block of shares of stock, looking for an opportune time to sell. We start by assuming that we are a small player which means it does not matter how many shares we sell, so we are going to assume that we have just one share. If we sell at time t, we receive a price that varies according to some random process over time, although we do not believe prices are trending up or down. Once we sell the stock, the process stops.

2.2 Basic model

2.2.1 State variables

Our process has two state variables: the "physical state," which captures whether or not we are still holding the asset, and an "informational state" which for this problem is the price of the stock.

Our "physical state" is given by

$$R_t^{asset} = \begin{cases} 1 & \text{if we are holding the stock at time } t, \\ 0 & \text{if we are no longer holding the stock at time } t. \end{cases}$$

If we sell the stock, we receive the price per share of p_t. This means our state variable is

$$S_t = (R_t^{asset}, p_t).$$

2.2.2 Decision variables

The decision variable is whether to hold or sell the stock. We write this using

$$x_t = \begin{cases} 1 & \text{if we sell the stock at time } t, \\ 0 & \text{if do not sell the stock at time } t. \end{cases}$$

We are only allowed to sell stock in this problem, so we have to obey the constraint

$$x_t \leq R_t^{asset}.$$

We are going to define our policy $X^\pi(S_t)$ which is going to define how we make decisions. At this stage, we introduce the notation for the policy, but defer designing the policy until later. This is what we mean when we say that we "model first, then solve."

2.2.3 Exogenous information

The only random process in our basic model is the change in price. There are two ways to write this. One is to assume that the exogenous information is the change in price. We can write this as

$$\hat{p}_{t+1} = p_{t+1} - p_t.$$

This means that our price process is evolving according to

$$p_{t+1} = p_t + \hat{p}_{t+1}.$$

We would then write our exogenous information W_{t+1} as

$$W_{t+1} = \hat{p}_{t+1}.$$

2.2.4 Transition function

The transition function consists of the equations that describe how the state evolves over time. The transition equation for R_t is given by

$$R_{t+1}^{asset} = R_t^{asset} - x_t, \tag{2.1}$$

where we have the constraint that $x_t \leq R_t^{asset}$ to ensure that we do not sell the asset when we no longer own it.

Next we have to write how the price process evolves over time. If we use the $\hat{p}_t$ notation, the transition function for the price p_t would be given by

$$p_{t+1} = p_t + \hat{p}_{t+1}. \tag{2.2}$$

Equations (2.1) and (2.2) make up what we call our *transition function* that we write as

$$S_{t+1} = S^M(S_t, X^\pi(S_t), W_{t+1}). \tag{2.3}$$

If we use our policy $X^\pi(S_t)$ to make decisions, and if we choose sample path ω that determines the sequence $W_1, W_2, \ldots, W_T$, then we can write a simulation of our process as

$$(S_0, x_0 = X^\pi(S_0), W_1(\omega), S_1, x_1 = X^\pi(S_1), W_2(\omega), \ldots, x_{T-1}, W_T(\omega), S_T).$$

Note that as we write the sequence, we index variables by their information content. For example, S_0 is an initial state, and x_0 depends only on S_0. By contrast, any variable indexed by t is allowed to "see" any of the outcomes of our exogenous process $W_1, \ldots, W_t$, but is not allowed to see W_{t+1}.

2.2.5 Objective function

We finish our model with a statement of our objective function, which then becomes the basis for evaluating policies. To start, we have to have some performance metric, which for this problem would be how much we earn from selling our stock. We can define a generic contribution function that we write as $C(S_t, x_t)$, which would be given by

$$C(S_t, x_t) = p_t x_t.$$

In our problem, $x_t = 0$ until we choose to sell. For now, assume that we are selling a single discrete asset (we could think of this as selling all of our shares at once). In this case, when we sell we would let $x_t = 1$, which is going to happen just once over our horizon. We write the dependence of $C(S_t, x_t)$ to capture the dependence on the state, which is because of the presence of the price p_t.

We now want to formulate our optimization problem. If the prices were given to us in advance, we would write

$$\max_{x_0,\ldots,x_{T-1}} \sum_{t=0}^{T-1} p_t x_t, \tag{2.4}$$

where we would impose the constraints

$$\sum_{t=0}^{T-1} x_t = 1, \tag{2.5}$$

$$x_t \leq 1, \tag{2.6}$$

$$x_t \geq 0. \tag{2.7}$$

This is fine when the problem is deterministic, but how do we model the problem to handle the uncertainty in the prices? What we do is to imagine that we are simulating a policy following a sample path ω of prices $p_1(\omega), p_2(\omega), \ldots$. Using a policy π, we would then generate a series of states using

$$S_{t+1}(\omega) = S^M(S_t(\omega), X^\pi(S_t(\omega)), W_{t+1}(\omega)).$$

We write $S_t(\omega)$ to express the dependence on the sample path. We could also have written $S_t^\pi(\omega)$ to express the dependence on the policy π, but we tend to suppress the dependence on the policy for simplicity.

If we follow policy π along this sample path, we can compute the performance using

$$\hat{F}^\pi(\omega) = \sum_{t=0}^{T-1} p_t(\omega) X^\pi(S_t(\omega)).$$

This is for one sample path. Note that we get a set of decisions $x_t(\omega)$ for each sample path from the policy

$$x_t(\omega) = X^\pi(S_t(\omega)).$$

This notation communicates that x_t is a random variable that depends on the sample path ω. For each sample path, we still get

$$\sum_{t=0}^{T-1} x_t(\omega) = 1,$$

which parallels our constraint above for the deterministic version of the problem. There is a time $\tau(\omega)$ which is the time where $x_t(\omega) = 1$ for $t = \tau(\omega)$. This time is known as a *stopping time* for this asset selling problem.

We can simulate over a sample of N samples $\omega^1, \ldots, \omega^n, \ldots, \omega^N$ and take an average using

$$\overline{F}^{\pi} = \frac{1}{N} \sum_{n=1}^{N} \hat{F}^{\pi}(\omega^n). \tag{2.8}$$

Finally, we write out the optimization problem in terms of finding the best policy, which we can write

$$\max_{\pi} \overline{F}^{\pi}. \tag{2.9}$$

We will see that the optimization problem stated by (2.9) where we would like to find the optimal policy is primarily aspirational. While we would certainly like the optimal policy, we will typically settle for the best policy that we can find (and can compute).

In practice we are typically using averages such as in Equation (2.8) for our optimization problem. However, this is just an approximation of taking an actual expectation, which we will write as

$$F^{\pi} = \mathbb{E}\hat{F}^{\pi} \approx \overline{F}^{\pi}. \tag{2.10}$$

By convention, when we write the expectation, we drop the indexing on ω and instead view $\hat{F}^{\pi}$ as a random variable, whereas $\hat{F}^{\pi}(\omega)$ is treated as a sample realization (this is standard notation in the stochastic modeling community, so just get used to it).

Using our expectation operator, we would write our objective function as

$$\max_{\pi} \mathbb{E}\hat{F}^{\pi}. \tag{2.11}$$

Often, we are going to write our objective function as

$$\max_{\pi} \mathbb{E}\left\{\sum_{t=0}^{T-1} p_t X^{\pi}(S_t)|S_0\right\}. \tag{2.12}$$

The form in Equation (2.11) (or (2.10) or (2.12)) is nice and compact. You just have to remember that it is almost never the case that we can actually compute the expectation, so we generally depend on running simulations and taking an average as we do in Equation (2.8).

Now we are left with the problem of searching over policies. We will always create our model first, and then turn to the problem of designing policies. Before we do this, we have to think about how we are going to model any uncertainties in S_0 and the exogenous information process $W_1, \ldots, W_T$.

2.3 Modeling uncertainty

We are going to need some way to sample observations of W_t which, for this problem, means modeling the evolution of prices p_t over time. One way is to draw samples from history. Imagine that we are interested in running our simulation over a period of a year. We could think of getting 10 years of past data, and scaling it so the processes all start at the same point. This is a popular and widely used method for creating samples, since it avoids making any simplifying modeling assumptions. The drawback is that this only gives us 10 sample paths, which may represent patterns from history that are no longer relevant.

The second strategy, which we will often use, is to estimate a statistical model. For our basic model, we might assume

$$p_{t+1} = p_t + \hat{p}_{t+1}, \tag{2.13}$$

where $\hat{p}_{t+1}$ is described by some probability distribution. A simple model would be to assume that $\hat{p}_{t+1}$ is normally distributed with mean 0 and variance σ^2. We might also start by assuming that the changes in prices $\hat{p}_t$ and $\hat{p}_{t+1}$ are independent, and that $\hat{p}_{t+1}$ is independent of the current price p_t (the latter assumption is a bit strong, but it will help us get started).

Table 2.1: Ten uniform random variables U, and ten corresponding samples of normally distributed price changes $\hat{p}$ with mean 0 and variance 1.

U	$\hat{p}$
0.8287	0.9491
0.6257	0.3206
0.9343	1.5086
0.4879	-0.0303
0.3736	-0.3223
0.8145	0.8947
0.0385	-1.7685
0.0089	-2.3698
0.9430	1.5808
0.3693	-0.3336

Most computer languages have functions for simulating observations from a normal distribution. For example, Excel provides the function $\texttt{Norm.inv}(p, \mu, \sigma)$, which returns the value w of a random variable W with mean μ and standard deviation σ where $P[W \leq w] = p$. A standard trick is to set $p = Rand()$, where $Rand()$ is an Excel function that returns a random variable that is uniformly distributed between 0 and 1. We can then write

$$\hat{p}_{t+1} = \texttt{Norm.inv}(Rand(), 0, \sigma),$$

which will give us a random observation of $\hat{p}_{t+1}$ that is normally distributed with mean 0 and standard deviation σ.

Table 2.1 illustrates ten observations of random variables U that are uniformly distributed between 0 and 1, and the corresponding samples of normally distributed price changes $\hat{p}$ with mean 0 and variance 1.

Equation (2.13) is a pretty basic price model, but it will help illustrate our modeling framework. Below, we are going to introduce some extensions which include a richer model.

2.4 Designing policies

We can envision several different policies for this problem. For example, a simple policy might be to sell if the price drops below some limit point that we think suggests that it is starting a big decline. Thus, we could write this policy as

$$X^{sell-low}(S_t|\theta^{low}) = \begin{cases} 1 & \text{if } p_t < \theta^{low}, \\ 1 & \text{if } t = T, \\ 0 & \text{otherwise.} \end{cases} \tag{2.14}$$

Another policy might be a "high-low" selling policy, where we want to sell if the price jumps too high or too low. Let $\theta^{high-low} = (\theta^{low}, \theta^{high})$. This might be written

$$X^{high-low}(S_t|\theta^{high-low}) = \begin{cases} 1 & \text{if } p_t < \theta^{low} \text{ or } p_t > \theta^{high}, \\ 1 & \text{if } t = T, \\ 0 & \text{otherwise.} \end{cases} \tag{2.15}$$

A possible objection to this policy could be that it prematurely sells a rising stock. Perhaps we just want to sell when the stock rises above a tracking signal. To handle this issue, first create a smoothed estimate of the price using

$$\bar{p}_t = (1-\alpha)\bar{p}_{t-1} + \alpha \hat{p}_t. \tag{2.16}$$

Now consider a tracking policy that we might write as

$$X^{track}(S_t|\theta^{track}) = \begin{cases} 1 & \text{if } p_t \geq \bar{p}_t + \theta^{track}, \\ 1 & \text{if } t = T, \\ 0 & \text{otherwise.} \end{cases} \tag{2.17}$$

In all the cases, we can only sell the asset (that is, $X^{track}(S_t|\theta^{track}) = 1$) if we are still holding the asset (which means $R_t^{asset} = 1$).

For this policy, we are going to need to tweak our model, because we now need $\bar{p}_t$ in order to make a decision. This means we would now write our state as

$$S_t = (R_t^{asset}, p_t, \bar{p}_t).$$

We can write our classes of policies as the set $\mathcal{F} = \{$"sell-low," "high-low,""track"$\}$. For each of these classes, we have a set of parameters that we can write as θ^f for $f \in \mathcal{F}$. For the "sell-low" and "track" policies there is a single parameter, while $\theta^{high-low}$ has two parameters.

Now we can write our search over policies π in a more practical way as searching over function classes $f \in \mathcal{F}$, and then searching over parameters $\theta^f \in \Theta^f$, where Θ^f tells us the range of possible values (capturing at the same time the dimensionality of θ^f).

The way that we designed the policies in this section may seem somewhat ad hoc, but this in fact is precisely how many policies are designed (including buy-sell strategies used by major hedge funds). Many types of policies are possible, some of which will perform better than others; we focus on these as illustrative examples. There is an art to designing policies which parallels the art of designing statistical models for estimation.

2.5 Policy evaluation

We indicated above that we can evaluate a policy by simulating it using

$$\hat{F}^\pi(\omega) = \sum_{t=0}^{T-1} p_t(\omega) X^\pi(S_t(\omega)),$$

where ω is used to represent a sample path of realizations of whatever exogenous random variables are used in the model. Table 2.2 illustrates a series of sample paths of prices. For example, imagine that we are using the "sell-low" policy with $\theta^{sell-low} = \$42$. Now consider testing it on sample path ω^5. The result would be

$$\hat{F}^{sell-low}(\omega^5) = \$41.53,$$

since \$41.53 is the first price that falls below \$42. If none of the prices fall below our sell point, then all of our policies are designed to sell at the end.

We can then evaluate each policy (both the class of policy, and the parameters for that class) by doing repeated simulations and taking an average. We write this as

$$\overline{F}^\pi = \frac{1}{N} \sum_{n=1}^{N} \hat{F}^\pi(\omega^n).$$

Sometimes we need to express a confidence interval, since $\overline{F}^\pi$ is nothing more than a statistical estimate. We would first compute an

Table 2.2: Illustration of a set of price paths.

	$t = 1$	$t = 2$	$t = 3$	$t = 4$	$t = 5$	$t = 6$	$t = 7$	$t = 8$
ω^n	p_1	p_2	p_3	p_4	p_5	p_6	p_7	p_8
ω^1	42.67	45.53	47.07	47.56	47.80	48.43	46.93	46.57
ω^2	46.35	43.15	42.51	40.51	41.50	41.00	39.16	41.11
ω^3	43.17	45.16	45.37	44.30	45.35	47.23	47.35	46.30
ω^4	45.24	45.67	46.18	46.22	45.69	44.24	43.77	43.57
ω^5	47.68	46.32	46.14	41.53	44.84	45.17	44.92	46.09
ω^6	47.83	44.70	43.05	43.77	42.61	44.32	44.16	45.29
ω^7	45.11	43.67	43.14	44.78	43.12	42.36	41.60	40.83
ω^8	46.78	44.98	44.53	45.42	46.43	47.67	43.68	49.03
ω^9	43.16	44.57	45.99	47.38	45.51	46.27	46.02	45.09
ω^{10}	46.57	45.01	46.73	42.08	47.40	49.14	49.03	48.74

estimate of the variance of our random variable $\hat{F}^\pi$, which we do using

$$(\hat{\sigma}^\pi)^2 = \frac{1}{N-1}\sum_{n=1}^{N}(\hat{F}^\pi(\omega^n) - \overline{F}^\pi)^2.$$

We then get our estimate of the variance of our average $\overline{F}^\pi$ using

$$(\bar{\sigma}^\pi)^2 = \frac{1}{N}(\hat{\sigma}^\pi)^2.$$

From this, we can construct a confidence interval to compare two policies which we might call π^A and π^B. Let μ^π be the true performance of policy π, where $\overline{F}^\pi$ is our statistical estimate of μ^π. We would like to get a confidence interval for the difference $\mu^{\pi^A} - \mu^{\pi^B}$. Our best estimate of this difference is $(\overline{F}^{\pi^A} - \overline{F}^{\pi^B})$. The variance of this difference is

$$Var(\overline{F}^{\pi^A} - \overline{F}^{\pi^B}) = (\bar{\sigma}^{\pi^A})^2 + (\bar{\sigma}^{\pi^B})^2,$$

where we assume that the estimates $\overline{F}^{\pi^A}$ and $\overline{F}^{\pi^B}$ are independent, which means that each policy is being tested on a different random sample of prices. When this is the case, we would compute our confidence interval using

$$\mu^{\pi^A} - \mu^{\pi^B} \in \left(\overline{F}^{\pi^A} - \overline{F}^{\pi^B} \pm z_\alpha\sqrt{(\bar{\sigma}^{\pi^A})^2 + (\bar{\sigma}^{\pi^B})^2}\right),$$

where z_α is the value of z such that a normally distributed random variable Z is greater than z with probability α. For example, $z_{.05} = 1.645$, which means $Prob[Z \geq 1.645] = .05$.

A better approach is to use the same samples to evaluate each policy. For example, we could test each policy on the same sample path ω chosen from Table 2.2. Testing our policies this way, we would get $\hat{F}^{\pi^A}(\omega)$ and $\hat{F}^{\pi^B}(\omega)$ (using the same set of prices $p_t(\omega)$), and then compute the difference

$$\delta\hat{F}^{A-B}(\omega) = \hat{F}^{\pi^A}(\omega) - \hat{F}^{\pi^B}(\omega).$$

Now we compute the average difference

$$\delta\overline{F}^{A-B} = \frac{1}{N}\sum_{n=1}^{N}\delta\hat{F}^{A-B}(\omega^n),$$

and the variance

$$(\delta\bar{\sigma}^{A-B})^2 = \frac{1}{N}\left(\frac{1}{N-1}\sum_{n=1}^{N}(\delta\hat{F}^{A-B}(\omega^n) - \delta\overline{F}^{A-B})^2\right).$$

Note that the variance $(\delta\bar{\sigma}^{A-B})^2$ will be smaller than the variance when independent samples are used. The confidence interval for the difference would then be

$$\delta\mu^{A-B} \in (\delta\overline{F}^{A-B} - z_\alpha\sqrt{\delta\bar{\sigma}^{A-B,2}}, \delta\overline{F}^{A-B} + z_\alpha\sqrt{\delta\bar{\sigma}^{A-B,2}}).$$

Computing confidence intervals may be useful when comparing different classes of policies. Alternatively, we may be comparing the performance of two physical designs (e.g. the speed of two machines or the locations of a facility). The choice of policy closely parallels any design decision for a system.

2.6 Extensions

2.6.1 Time series price processes

Imagine that we want a somewhat more realistic price process that captures autocorrelation over time. We might propose that

$$p_{t+1} = \eta_0 p_t + \eta_1 p_{t-1} + \eta_2 p_{t-2} + \varepsilon_{t+1}, \tag{2.18}$$

where we still assume that the random noise ε_t is independent (and identically distributed) over time. We are also going to assume for the moment that we know the coefficients $\eta = (\eta_0, \eta_1, \eta_2)$.

This price model requires a subtle change in our model, specifically the state variable. We are going to replace our old transition equation for prices, (2.2), with our new time series model given in (2.18). To compute p_{t+1}, it is no longer enough to know p_t, we now also need to know p_{t-1} and p_{t-2}. Our state variable would now be given by

$$S_t = (R_t, p_t, p_{t-1}, p_{t-2}).$$

For the policies we have considered above, this does not complicate our model very much. Later, we are going to introduce policies where the additional variables represent a major complication.

2.6.2 Time series price process with learning

Now assume that our time series price process is given by

$$p_{t+1} = \bar{\eta}_{t0} p_t + \bar{\eta}_{t1} p_{t-1} + \varepsilon_{t+1}, \tag{2.19}$$

where $\varepsilon \sim N(0, 4^2)$ and where $\bar{\eta}_t = (\bar{\eta}_{t0}, \bar{\eta}_{t1})$ is our *estimate* of η given what we know at time t (in the previous section, we assumed θ was known).

There are simple formulas that govern the updating of $\bar{\eta}_t$ to $\bar{\eta}_{t+1}$ given our estimates $\bar{\eta}_t$ and the observation of the next price p_{t+1}.

We first let

$$\bar{p}_t(p_t|\bar{\eta}_t) = \bar{\eta}_{t0} p_t + \bar{\eta}_{t1} p_{t-1}$$

be our estimate of p_{t+1} given what we know at time t. The error in this estimate is given by

$$\hat{\varepsilon}_{t+1} = \bar{p}(p_t|\bar{\eta}_t) - p_{t+1}. \tag{2.20}$$

Now let the vector ϕ_t be the vector of explanatory variables in our price process which is given by

$$\phi_t = \begin{pmatrix} p_t \\ p_{t-1} \end{pmatrix}.$$

Next we define the 2×2 matrix M_t which is updated recursively using

$$M_t = M_{t-1} - \frac{1}{\gamma_t}(M_{t-1}\phi_t(\phi_t)^T M_{t-1}), \tag{2.21}$$

where γ_t is a scalar computed using

$$\gamma_t = 1 + (\phi_t)^T M_{t-1}\phi_t. \tag{2.22}$$

We can now update $\bar{\eta}_t$ using

$$\bar{\eta}_{t+1} = \bar{\eta}_t - \frac{1}{\gamma_t} M_t \phi_t \hat{\varepsilon}_t. \tag{2.23}$$

Exercise 2.6 will dive into these equations further.

2.6.3 Basket of assets

Another twist arises when we are considering a basket of assets. Let p_{ti} be the price of asset i. Assume for the moment that each price evolves according to the basic process

$$p_{t+1,i} = p_{ti} + \varepsilon_{t+1,i}.$$

We could assume that the noise terms $\varepsilon_{t+1,i}$ are independent across the assets $i \in \mathcal{I}$, but a more realistic model would be to assume that the prices of different assets are correlated. Let

$$\sigma_{ij} = Cov_t(p_{t+1,i}, p_{t+1,j})$$

be the covariance of the random prices $p_{t+1,i}$ and $p_{t+1,j}$ for assets i and j given what we know at time t. Assume for the moment that we know the covariance matrix Σ, perhaps by using a historical dataset to estimate it (but holding it fixed once it is estimated).

We can use the covariance matrix to generate sample realizations of correlated prices using a technique called Cholesky decomposition. It proceeds by creating what we call the "square root" of the covariance matrix Σ which we store in a lower triangular matrix L. In python, using the NumPy package, we would use the python command

$$L = \texttt{scipy.linalg.cholesky}(\Sigma, lower = True).$$

The matrix L allows us to obtain the matrix Σ using

$$\Sigma = L^T L.$$

Now let Z be a vector of random variables, one for each asset, where $Z_i \sim N(0,1)$ (virtually every programming language has routines for creating random samples from normal distributions with mean 0, variance 1). Let p_t, p_{t+1} and Z be column vectors (dimensioned by the number of assets). We first create a sample $\hat{Z}$ by sampling from $N(0,1)$ $|\mathcal{I}|$ times. Our sample of prices p_{t+1} is then given by

$$p_{t+1} = p_t + L\hat{Z}.$$

For this problem, our state variable is given by

$$S_t = (R_t, p_t)$$

where $R_t = (R_{ti})_{i \in \mathcal{I}}$ captures how many shares of each asset we own, while p_t is our current vector of prices. The covariance matrix Σ is not in the state variable because we assume it is static (which means we put it in S_0). The answer changes if we were to update the covariance matrix with each new observation, in which case we would write the covariance matrix as Σ_t to capture its dependence on time. Since it now varies dynamically, the state variable would be

$$S_t = (R_t, p_t, \Sigma_t).$$

2.7 What did we learn?

We used this problem to illustrate different types of PFA policies:

- We used a simple asset selling problem to illustrate a sequential decision problem with both a physical state (whether we are holding an asset, or how much), and the price we could sell it at time t.

- We showed how to model uncertainty and introduce the notion of a sample path ω for the exogenous information process $W_1, \ldots, W_T$.

- We illustrated several simple policies in the PFA class.

- We showed how to simulate a policy.
- We introduced some complexity in the price process (where the price p_{t+1} depends on recent history of prices), and the situation where the selling price depends on a basket of assets, which is a more complex, multidimensional information process. Note that this does not introduce any significant complexity other than modeling the process. It creates a much higher dimensional state variable, but that does not represent a significant form of complexity for the problem of designing or evaluating policies (this is not true of all classes of policies).
- We showed how to update a linear model of the price process.

2.8 Exercises

Review questions

2.1 — We write the transition function as

$$S_{t+1} = S^M(S_t, x_t, W_{t+1}). \tag{2.24}$$

a) Why do we write the exogenous information as W_{t+1} rather than W_t?

b) At what point in time would we be computing S_{t+1} using Equation (2.24)?

2.2 — What is the difference between p_t and $p_t(\omega)$?

2.3 — As you test out different policies, does the structure of the transition function change?

2.4 — The objective function in Equation (2.4) is written for a deterministic version of the problem. If we solve this optimization problem, does the decision x_t at time t depend on the prices $p_{t'}$ for $t' > t$?

Problem solving questions

2.5 — Using the prices in Table 2.2, use the policy that you will sell when the price falls below \$44.00. Compute the objective function $\hat{F}(\omega^n)$ for $n = 1, \ldots, 10$. Compute the average selling price and its variance.

2.6 — The questions below walk you through the steps of modeling the selling of an asset (say, a single share of a stock).

a) Assume that you are simulating your prices using data from a mathematical model given by

$$p_{t+1} = \eta_0 p_t + \eta_1 p_{t-1} + \varepsilon_{t+1}, \tag{2.25}$$

where $\varepsilon \sim N(0, 6^2)$.

We do not know the value of $\eta = (\eta_0, \eta_1)$, but our belief about the true value of η is that it is multivariate normal, with $\eta \sim MVN(\bar{\eta}, \Sigma)$ where

$$\bar{\eta}_t = \begin{bmatrix} .7 \\ .3 \end{bmatrix}.$$

Assume that the covariance matrix Σ_t is given by

$$\Sigma_t = \begin{bmatrix} (.2)^2 & (.05)^2 \\ (.05)^2 & (.1)^2 \end{bmatrix}.$$

where $\Sigma_{tij} = Cov(\eta_i, \eta_j)$ for $i, j \in (0, 1)$. Assume that at time t, $p_t = 20$, $p_{t-1} = 24$ and we observe $p_{t+1} = 18.2$.

Using the equations in Section 2.6.2, what is the updated estimates of $\bar{\eta}_{t+1}$ and Σ_{t+1}? Give the updating equation and compute $\bar{\eta}_{t+1}$ and Σ_{t+1} numerically.

b) What is the state variable for this problem? Give the list of variables. Note that you may have to add variables as you progress through the exercise (you do not have all the information at this point). Make sure you include all the information you need to update $\bar{\eta}_{t+1}$ from $\bar{\eta}_t$. How many dimensions does your state variable have (this is the same as asking how many variables are in S_t).

c) Our trader makes trading decisions based on a 7-day moving average of prices. Assume we are on day t, and the 7-day moving average is computed using

$$\bar{p}_t = \frac{1}{7} \sum_{t'=t-7+1}^{t} p_{t'}. \tag{2.26}$$

The trader will sell an asset if $p_t < \bar{p}_t - \theta^{sell}$. Write out the decision rule as a policy $X^\pi(S_t|\theta^{sell})$ that returns 1 if we sell the asset and 0 otherwise.

d) What is the exogenous information?

e) Write out the transition equations. You need an equation for each element of S_t.

f) Write out the objective function (and remember to use an expectation operator for each random variable as described in the instructions). Be sure to specify what you are optimizing over given the class of policy specified in part (c). Assume that you are training your policy on historical data in an offline setting.

2.7 — You need to run a simulation of electricity prices, which are notoriously heavy-tailed. You collect the data shown in Table 2.3.

Table 2.3: Set of prices for exercise 2.7.

Time	Price
1	20
2	32
3	26
4	180
5	30
6	45
7	18
8	120
9	57
10	15

a) Use the data in the table to produce a cumulative distribution function. You will need to plot the cdf, which looks like a step function.

b) You now observe three realizations of prices: 25, 18, 160. Use the cumulative distribution function from (a) to create three realizations of a random variable that is uniformly distributed between 0 and 1 (call this random variable U).

c) Now use these observations of U to create three observations of a random variable Z that is normally distributed with mean 0, variance 1. Be sure that your method for generating these random variables is clear. [Hint: use the cumulative distribution for a normal (0,1) random variable just as you used the cdf you created in part (a) to create uniform random variables in part (b).]

d) Assume that you used this method for creating a sequence of normal (0,1) random variables to fit a linear model of the form

$$Z_{t+1} = .7Z_t + .3Z_{t-1} + \varepsilon_{t+1} \tag{2.27}$$

where ε_{t+1} is normally distributed with mean 0 and variance 1 (we know this because we are simulating normal (0,1) random variables). Starting with $t = 1$ and $Z_0 = 0.5$ and $Z_1 = -0.3$, generate Z_2, Z_3, Z_4 assuming that $\varepsilon_2 = -.6, \varepsilon_3 = 2.2, \varepsilon_4 = 1.4$. Then, use your cumulative distribution from part (a) to generate observations of P_2, P_3 and P_4.

Programming questions

These exercises use the Python module *AssetSelling*.[1]

2.8 — Our basic "high-low" selling policy was given by

$$X^{high-low}(S_t|\theta^{high-low}) = \begin{cases} 1 & \text{if } p_t < \theta^{low} \text{ or } p_t > \theta^{high}, \\ 1 & \text{if } t = T, \\ 0 & \text{otherwise.} \end{cases} \tag{2.28}$$

In addition to the module *AssetSelling*, you will also need the spreadsheet

[1]See http://tinyurl.com/sdagithub

"Chapter2_asset_selling_policy"[2] which provides parameters to be used by the python module.

a) Simulate the policy for 200 time periods using the parameters:

$$\begin{aligned} \theta^{min} &= 6, \\ \theta^{max} &= 13, \\ T &= 20. \end{aligned}$$

b) Perform a search for the best value of θ^{min} and θ^{max} by fixing $\theta^{max} = 13$, and then searching in increments of 1 for the best θ^{min}. Then fix at that value of θ^{min} and perform a similar search for θ^{max} (enforce the constraint that $\theta^{max} = \theta^{min} + 2$).

2.9 — Consider a policy that understands that the price of the asset might be rising, which means that static buy-sell limits might not be effective. Assume that we forecast the price for time $t+1$ using the fitted time series model

$$\bar{p}_t = 0.7p_t + 0.2p_{t-1} + 0.1p_{t-2}.$$

We are going to use $\bar{p}_t$ as a forecast of p_{t+1} given what we know at time t.

a) What is the state variable for this problem? If prices are discretized to the nearest 0.1, and prices are assumed to range between 0 and 100, what is the size of the state space?

b) Assume we introduce the policy

$$X^{time-series}(S_t|\theta^{low}) = \begin{cases} 1 & \text{if } p_t < \bar{p}_t - \theta \text{ or } p_t > \bar{p}_t + \theta, \\ 1 & \text{if } t = T, \\ 0 & \text{otherwise.} \end{cases} \tag{2.29}$$

In this version of the policy, we are looking for sudden deviations from the expected price. Plot the graph of the objective function (contribution) vs. θ. Comment on your plot. Note that while the state space is quite large, there is no change in the complexity of finding the best policy in this class.

[2] See https://tinyurl.com/sdamodelingsupplements

(Hint: You would need to change the state variable, change the high-low policy, and create an outer loop over different values of θ in the module *DriverScript*. You are encouraged to play around with your code and choose your own ranges of values. Include how long it took you to run your code with your choice of values.)

2.10 — (Continuing from Exercise 2.9) Imagine that our stock follows a seasonal pattern, suggesting that our buy-sell signals should be time-dependent. This means that we need to replace θ with θ_t. Discuss how this would complicate our policy search process.

2.11 — (Continuing from Exercise 2.9) We might next feel that our buy-sell signal should depend on the price. For example, if prices are higher, we might feel that we look for a larger deviation from $\bar{p}_t$ in Equation (2.29) than if the prices were smaller. This means that we would replace the constant vector θ with a function $\theta(\bar{p}_t)$.

a) Describe a lookup table representation of $\theta(\bar{p}_t)$, and draw a graph depicting how you think this function might look. This would require discretizing $\bar{p}_t$ into, say, 10 ranges. How does this complicate the problem of searching for policies?

b) Suggest a parametric form for $\theta(\bar{p}_t)$ that captures your intuition that θ should be larger if $\bar{p}_t$ is larger? What parameters do you now have to search over?

3

Adaptive Market Planning

3.1 Narrative

There is a broad class of problems that involve allocating some resource to meet an uncertain (and sometimes unobservable) demand. Examples include:

- Stocking a perishable inventory (e.g. fresh fish) to meet a demand where leftover inventory cannot be held for the future.
- Stocking parts for high-technology manufacturing (e.g. jet engines) where we need to order parts to meet known demand, but where the parts may not meet required specifications and have to be discarded. So, we may need to order eight parts to meet the demand of five because several of the parts may not meet required engineering specifications.
- We have to allocate time to complete a task (such as driving to work, or allocating time to complete a project).
- We have to allocate annual budgets for activities such as marketing. Left-over funds are returned to the company.

The simplest problem involves making these decisions to meet an uncertain demand with a known distribution, but the most common applications involve distributions that are unknown and need to be learned. There may be other information such as the availability of forecasts of the demand, as well as dynamic information such as the market price for the fresh fish (which may be known or unknown before the resource decision is made).

This problem has been widely studied since the 1950's, originally known as the "single period inventory problem" but currently is primarily identified as the "newsvendor problem." It is widely used as the canonical problem in optimization under uncertainty.

The newsvendor problem is typically formulated as

$$\max_x \mathbb{E}F(x, W) = \mathbb{E}(p \min\{x, W\} - cx), \tag{3.1}$$

where x is our decision variable that determines the amount of resource to meet the demand, and where W is the uncertain demand for the resource. We assume that we "purchase" our resource at a unit cost of c, and we sell the smaller of x and W at a price p (which we assume is greater than c). The objective function given in Equation (3.1) is called the *asymptotic* form of the newsvendor problem.

If W was deterministic (and if $p > c$), then the solution is easily verified to be $x = W$. Now imagine that W is a random variable with probability distribution $f^W(w)$ (W may be discrete or continuous). Let $F^W(w) = Prob[W \leq w]$ be the cumulative distribution of W. If W is continuous, and if we could compute $F(x) = \mathbb{E}F(x, W)$, then the optimal solution x^* would satisfy

$$\left.\frac{dF(x)}{dx}\right|_{x=x^*} = 0.$$

Now consider what is known as the *stochastic gradient*, where we take the derivative of $F(x, W)$ assuming we know W, which is given by

$$\frac{dF(x, W)}{dx} = \begin{cases} p - c & x \leq W, \\ -c & x > W. \end{cases} \tag{3.2}$$

This is a gradient (that is, a derivative) of $F(x, W)$ given the random variable W, which is "stochastic" because it depends on the random variable W which is revealed only after we choose x. This is the reason that $dF(x,W)/dx$ in Equation (3.2) is called a "stochastic gradient."

Taking expectations of both sides of (3.2) gives

$$\begin{aligned}\mathbb{E}\frac{dF(x,W)}{dx} &= (p-c)Prob[x \leq W] - cProb[x > W] \\ &= (p-c)(1-F^W(x)) - cF^W(x) \\ &= (p-c) - pF^W(x) \\ &= 0 \quad \text{for } x = x^*.\end{aligned}$$

We can now solve for $F^W(x^*)$ giving

$$F^W(x^*) = \frac{p-c}{p}.$$

Thus, as c decreases to 0, we want to order an amount x^* that will satisfy demand with probability 1. As c approaches p, then the optimal order quantity will satisfy demand with a probability approaching 0.

This means that we compute $(p-c)/p$, which is a number between 0 and 1, and then find the quantity x^* that corresponds to the order quantity where the probability that the random demand is less than x^* is equal to $(p-c)/p$.

We have just seen two situations where we can find the order quantity exactly: when we know W in advance (we might call this the perfect forecast) or when we know the distribution of W. This result has been known since the 1950's, sparking a number of papers for estimating the distribution of W from observed data, handling the situation where we cannot observe W directly when $x < W$ (that is, we only observe sales, rather than demand, a situation known as "censored demands").

We are going to tackle the problem where the demand distribution is unknown. Our approach will be to use a sequential search algorithm given by

$$x^{n+1} = \max\left\{0, x^n + \alpha_n \left.\frac{dF(x,W^{n+1})}{dx}\right|_{x=x^n}\right\}, \tag{3.3}$$

where α_n is known as a *stepsize*. Our challenge will be to choose α_n at each iteration.

3.2 Basic model

3.2.1 State variables

The state variable captures the information we have at time n that we need, along with the policy and the exogenous information, to compute the state at time $n+1$. For our search procedure in Equation (3.3), our state variable is given by

$$S^n = (x^n).$$

3.2.2 Decision variables

The trick with this problem is recognizing the decision variable. It is tempting to think that x^n is the decision, but in the context of this algorithm, the real decision is the stepsize α_n. As with all of our sequential decision problems, the decision (that is, the stepsize) is determined by what is typically referred to as a stepsize rule, but is sometimes called a stepsize policy that we denote by $\alpha^\pi(S^n)$.

Normally we introduce policies later, but to help with the understanding the model, we are going to start with a basic stepsize policy called a *harmonic stepsize rule* given by

$$\alpha^{harmonic}(S^n|\theta^{step}) = \frac{\theta^{step}}{\theta^{step}+n-1}.$$

This is a simple deterministic stepsize rule, which means that we know in advance the stepsize α_n once we know n. Below we introduce a more interesting stochastic stepsize policy that requires a richer state variable.

We are also going to let $X^\pi(S^n)$ be the value of x^n determined by stepsize policy $\alpha^\pi(S^n)$.

3.2.3 Exogenous information

The exogenous information is the random demand W^{n+1} for the resource (product, time or money) that we are trying to meet with our supply of product x^n. We may assume that we observe W^{n+1} directly, or we may just observe whether $x^n \leq W^{n+1}$, or $x^n > W^{n+1}$.

3.2.4 Transition function

The transition equation, for the setting where x is unconstrained, is given by

$$x^{n+1} = x^n + \alpha_n \nabla_x F(x^n, W^{n+1}). \tag{3.4}$$

We note that it is possible that Equation (3.4) produces a value $x^{n+1} < 0$, which cannot be implemented. The fix here is simple: just set $x^{n+1} = 0$.

3.2.5 Objective function

At each iteration we receive a net benefit given by

$$F(x^n, W^{n+1}) = p \min\{x^n, W^{n+1}\} - cx^n.$$

Now we have to construct an objective function to find the best policy. We can approach this problem setting in two ways. In the first, we assume that we have to learn in the field, while the second assumes that we have access to a simulator for learning the policy.

Optimizing in the field

If we are experiencing our decisions in the field, we want to maximize the *cumulative reward* over some horizon. This means we need to find the best policy (which in this setting means the best stepsize rule) by solving

$$\max_\pi \mathbb{E}\left\{\sum_{n=0}^{N-1} F(X^\pi(S^n|\theta), W^{n+1})|S^0\right\}. \tag{3.5}$$

where $S^{n+1} = S^M(S^n, X^\pi(S^n), W^{n+1})$ describes the evolution of the algorithm (for example, the transition function given by Equation (3.4)). Here, π refers to the type of stepsize rule (we consider several below), and any tunable parameters (such as θ^{step}).

Oddly, the story behind the newsvendor problem always involves learning in the field, and yet the cumulative reward in Equation (3.5) is never used as the objective function. We mention this for readers who do a literature search on the "newsvendor problem."

Optimizing using a simulator

Alternatively, we might be using a simulator where we are going to run our search for N iterations, ending with x^N. We are going to rename this final solution $x^{\pi,N}$ to express the dependence on the stepsize policy $\alpha^\pi(S^n)$.

Our final solution $x^{\pi,N}$ is a random variable since it depends on the sequence $W^1,\ldots,W^n$. As before, we are going to let ω represent a sample realization of $W^1(\omega),\ldots,W^n(\omega)$, and we write our solution as $x^{\pi,N}(\omega)$ to indicate that this is the solution we obtained when we used the sample path ω.

Since we are using a simulator, we only care about the performance of the final solution (also called the *final reward*), which we write as

$$F(x^{\pi,N},\widehat{W}) \quad = \quad p\min\{x^{\pi,N},\widehat{W}\} - cx^{\pi,N}, \tag{3.6}$$

where $\widehat{W}$ is a random variable that we use for testing the performance of $x^{\pi,N}$.

This means we have two random variables in our objective function given in (3.6). For a single set of realizations of $W^1(\omega),\ldots,W^n(\omega)$, we get a solution $x^{\pi,N}(\omega)$. Now let ψ be a sample realization of $\widehat{W}$. So, if we have a sample realization of the solution $x^{\pi,N}(\omega)$, and a sample realization of our testing variable $\widehat{W}(\psi)$, our performance would be

$$F(x^{\pi,N}(\omega),\widehat{W}(\psi)) \quad = \quad p\min\{x^{\pi,N}(\omega),\widehat{W}(\psi)\} - cx^{\pi,N}(\omega). \tag{3.7}$$

What we really want to do is to take averages over the possible realizations of both $x^{\pi,N}(\omega)$ and $\widehat{W}(\psi)$, which we can write using

$$\overline{F}^\pi \quad = \quad \frac{1}{N}\frac{1}{M}\sum_{\omega=1}^{N}\sum_{\psi=1}^{M} p\min\{x^{\pi,N}(\omega^n),\widehat{W}(\psi^m)\} - cx^{\pi,N}(\omega^n). \tag{3.8}$$

The estimate $\overline{F}^\pi$ represents an average over N samples of the sequence $W^1(\omega),\ldots,W^n(\omega)$, and M samples of the testing variable $\widehat{W}(\psi)$.

3.3 Uncertainty modeling

Let $f^W(w)$ be the distribution of W (this might be discrete or continuous), with cumulative distribution function $F^W(w) = Prob[W \leq w]$.

We might assume that the distribution is known with an unknown parameter. For example, imagine that W follows a Poisson distribution with mean μ given by

$$f^W(w) = \frac{\mu^w e^{-\mu}}{w!}, \quad w = 0, 1, 2, \ldots.$$

We may assume that we know μ, in which case we could solve this problem using the analytical solution given at the beginning of the chapter. Assume, instead, that μ is unknown, but with a known distribution

$$p_k^\mu = Prob[\mu = \mu_k].$$

Note that this distribution $p^\mu = (p_k^\mu)_{k=1}^K$ would be modeled in our initial state S^0.

3.4 Designing policies

We have already introduced two choices of stepsize policies which we write as $\alpha^\pi(S^n)$ to mimic our style elsewhere for writing policies.

A wide range of stepsize policies (often called stepsize rules) have been suggested in the literature. One of the simplest and most popular is the harmonic stepsize policy given by

$$\alpha^{harmonic}(S^n|\theta^{step}) = \frac{\theta^{step}}{\theta^{step} + n - 1}.$$

Figure 3.1 illustrates the behavior of the harmonic stepsize rule for different values of θ^{step}.

The harmonic stepsize policy is also known as a deterministic policy, because we know its value for a given n in advance. The challenge with deterministic policies is that they are not allowed to adapt to the data. For this reason, it is often useful to use a stochastic rule. One of the earliest and simplest examples is Kesten's rule

$$\alpha^{kesten}(S^n|\theta^{step}) = \frac{\theta^{step}}{\theta^{step} + K^n - 1},$$

where K^n is a counter that counts how many times the gradient has switched direction. We determine this by asking if the product (or inner product, if x is a vector) $(\nabla_x F(x^n, W^{n+1}))^T \nabla_x F(x^{n-1}, W^n) < 0$. If

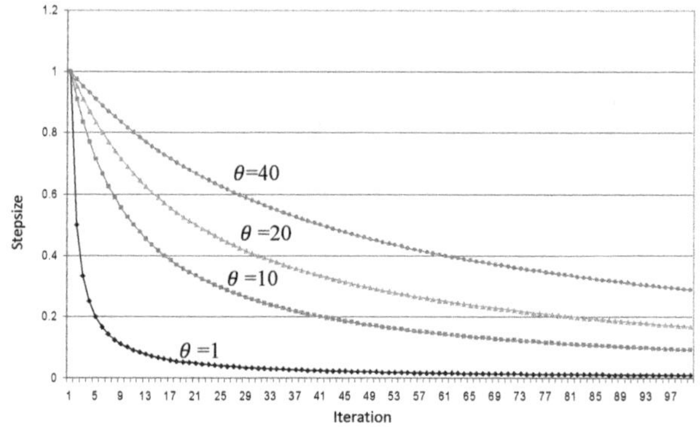

Figure 3.1: Harmonic stepsizes for different values of θ^{step}.

the gradient is switching directions, then it means that we are in the vicinity of the optimum and are stepping past it, so we need to reduce the stepsize. This formula is written

$$K^{n+1} = \begin{cases} K^n + 1 & \text{if } (\nabla_x F(x^n, W^{n+1}))^T \nabla_x F(x^{n-1}, W^n) < 0, \\ K^n & \text{otherwise.} \end{cases} \tag{3.9}$$

Now we have a stepsize that depends on a random variable K^n, which is why we call it a stochastic stepsize rule. If we use Kesten's rule, we have to modify our state variable to include K^n, giving us

$$S^n = (x^n, K^n).$$

We also have to add Equation (3.9) to our transition function.

Another stepsize rule, known as AdaGrad, is particularly well suited when x is a vector with element x_i, $i = 1, \ldots, I$. To simplify the notation a bit, let the stochastic gradient with respect to element x_i be given by

$$g_i^n = \nabla_{x_i} F(x^{n-1}, W^n).$$

Now create a $I \times I$ diagonal matrix G^n where the $(i,i)th$ element G_{ii}^n is given by

$$G_{ii}^n = \sum_{m=1}^{n} (g_i^n)^2.$$

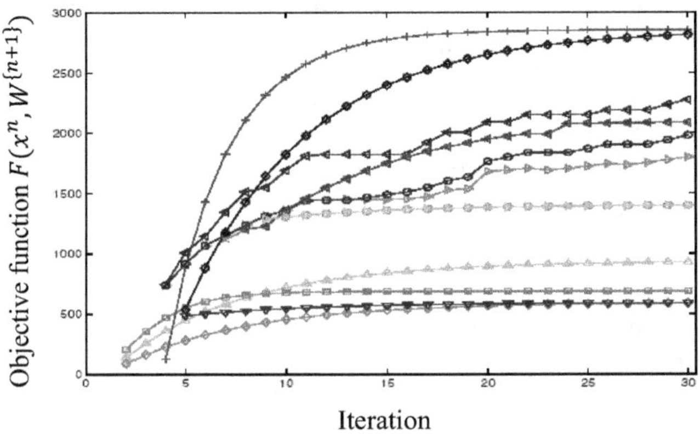

Figure 3.2: Plot of $F(x^n, W^{n+1})$ for different stepsize rules, illustrating different rates of convergence.

We then set a stepsize for the ith dimension using

$$\alpha_{ni} = \frac{\theta}{(G^n_{ii})^2 + \epsilon}, \tag{3.10}$$

where θ is a tunable parameter (comparable to θ^{step} in our harmonic stepsize formula) and ϵ is a small number (e.g. 10^{-8} to avoid the possibility of dividing by zero).

Figure 3.2 illustrates different rates of convergence for different stepsize rules, showing $F(x^n, W^{n+1})$ as a function of the number of iterations. If we were optimizing the final reward in (3.8), we could just pick the line that is highest, which depends on the budget N. If we are optimizing the cumulative reward in Equation (3.5), then we have to focus on the area under the curve, which favors rapid initial convergence.

3.5 Extensions

1) Imagine that we do not know μ, but let's assume that μ can take on one of the values $(\mu_1, \mu_2, \ldots, \mu_K)$. Let H^n be the history of observations up through the nth experiment, and let H^0 be the

initial empty history. We assume we start with an initial prior probability on μ that we write as

$$p_k^0 = Prob[\mu = \mu_k | H^0].$$

After we have observed $W^1, \ldots, W^n$, we would write our updated distribution as

$$p_k^n = Prob[\mu = \mu_k | H^n].$$

We can update $p^n = (p_k^n)_{k=1}^K$ using Bayes theorem

$$p_k^{n+1} = Prob[\mu = \mu_k | W^{n+1} = w, H^n] \tag{3.11}$$

$$= \frac{Prob[W^{n+1} = w | \mu = \mu_k, H^n] Prob[\mu = \mu_k | H^n]}{Prob[W^{n+1} = w | H^n]} \tag{3.12}$$

$$= \frac{Prob[W^{n+1} = w | \mu = \mu_k] p_k^n}{Prob[W^{n+1} = w | H^n]}, \tag{3.13}$$

where

$$Prob[W^{n+1} = w | H^n] = \sum_{k=1}^{K} Prob[W^{n+1} = w | \mu = \mu_k] p_k^n.$$

With this extension to the basic model, we have two probability distributions: the belief on the true mean μ, and the random demand W given μ. To include this extension, we would have to insert p^n into our state variable, so we would write

$$S^n = (x^n, p^n).$$

2) Imagine that our problem is to purchase a commodity (such as oil or natural gas) in month n to be used during month $n+1$. We purchase the commodity at a unit cost c, and sell it up to an unknown demand D^{n+1} at an unknown price p^{n+1}. We would write our objective for this problem as

$$F^n(x^n, W^{n+1}) = p^{n+1} \min(x^n, D^{n+1}) - cx^n. \tag{3.14}$$

3.6 What did we learn?

- We used the context of a newsvendor problem to illustrate a stochastic gradient algorithm as a sequential decision problem.

We showed how to model a stochastic gradient algorithm using the five elements of a sequential decision problem introduced in Chapter 1.

- We introduced several examples of PFA policies for choosing the stepsizes.
- The newsvendor problem is classically stated as a static problem where we search for the best solution where we are only interested in the performance of our final choice of x. In this chapter, we introduced two objectives: the *cumulative reward* for online learning (optimizing) in the field, and the *final reward* if we were using a simulator to design the best learning policy.
- We introduce the idea of using a probability distribution (in this case a Poisson distribution) for the random demands for product, where the mean of the Poisson distribution is itself a random variable.
- We introduce the extension of adaptively learning the probability distribution for the mean of the Poisson distribution.
- We also introduce the issue of making the decision x_t depend on other state variables such as the price p_t. This is the same as creating a policy $X^\pi(S_t)$ where the state S_t depends (in this setting) on the price p_t. This is a major shift in thinking about the newsvendor problem, but falls in the same class as all of our sequential decision problems.

3.7 Exercises

Review questions

3.1 — What is the decision variable for our sequential search algorithm?

3.2 — Give examples of searching over classes of policies (give two examples) and the tunable parameters for each class of policy.

3.3 — Write out what is meant by a *cumulative reward* objective and a *final reward* objective.

3.4 — When searching over the tunable parameter θ^{step} for the harmonic stepsize rule, how do you think the optimal value of θ^{step} obtained using a cumulative reward would compare to the optimal value when using a final reward?

3.5 — Assuming that we do not know the distribution of the demand W, argue why it does not make sense to find the optimal x^* in a simulator. Given this, it makes more sense to use the simulator to optimize the learning policy. If we use a simulator to optimize the learning policy, what objective function would be appropriate for this learning exercise?

Problem solving questions

3.6 — A major industrial gases company has to purchase contracts for electricity a month in advance using a "take or pay" contract. If the company contracts to purchase x_t megawatt-hours for month $t+1$, it pays a price p_t regardless of whether it needs the power or not. But if the load (demand) L_{t+1} in month $t+1$ exceeds x_t, then the company has to purchase power from the grid at a spot price p^{spot}_{t+1}. The cost of satisfying the load in month $t+1$ is then

$$C(S_t, W_{t+1}) = p_t x_t + p^{spot}_{t+1} \max\{0, L_{t+1} - x_t\}.$$

We are able to observe the different prices and loads, but we do not know their probability distribution. Our goal is to minimize costs over a year.

Assume that x_t is discrete with values $x_1, \ldots, x_M$. Let $(\bar{\mu}_{tx}, \beta_{tx})$ be the mean and precision of our estimate of $\mathbb{E}C(S_t, W_{t+1})$ and assume that we use a policy called *interval estimation*, $X^{IE}(S_t|\theta)$, to choose x_t:

$$X^{IE}(S_t|\theta^{IE}) = \arg\min_x \left(\bar{\mu}_{tx} - \theta^{IE}\sqrt{\frac{1}{\beta_{tx}}}\right). \tag{3.15}$$

a) Give the state variable S_t and exogenous information W_{t+1}.

b) Write the objective function for finding θ^{IE} to minimize cumulative costs over a year. Show the expectation over each random variable by writing the random variable as a subscript of the expectation operator (as in $\mathbb{E}_Y$). Then show how to write the expectation as a simulation assuming you have K samples of each random variable.

c) Give the formula for finding the gradient of the objective function in (b) using a numerical derivative, and write out a stochastic gradient algorithm for finding a good value of θ within N iterations.

d) Assume now that prices evolve according to

$$p_{t+1} = \eta_0 p_t + \eta_1 p_{t-1} + \eta_2 p_{t-2}.$$

What is the state variable now, and how does adding dimensions to the state variable complicate the problem of finding the optimal η above?

3.7 — Consider extension 2 in Section 3.5, where the price p now changes with iterations, and where the price we receive at time n, is not known at time n, so we designate it by p^{n+1}. For now, assume that p^{n+1} is independent of p^n, and that D^{n+1} is independent of D^n.

a) For the model in Equation (3.14), give the state variable S^n and the exogenous information variable W^n.

b) Give the stochastic gradient algorithm for this problem, and show that it is basically the same as the one when price was constant.

3.8 — Extend Exercise 3.7, but now assume that the prices evolve according to

$$p^{n+1} = \eta_0 p^n + \eta_1 p^{n-1} + \eta_2 p^{n-2} + \varepsilon^{n+1}$$

where ε^{n+1} is a mean 0 noise term that is independent of the price process.

a) For the model in Equation (3.14), give the state variable S^n and the exogenous information variable W^n.

b) Give the stochastic gradient algorithm for this problem.

3.9 — Now assume that our objective is to optimize

$$F^n(x^n, W^{n+1}) \quad = \quad p^n \min(x^n, D^{n+1}) - cx^n. \tag{3.16}$$

The only difference between Equations (3.16) and (3.14) is that now we get to see the price p^n *before* we choose our decision x^n. We know this by how the price is indexed.

a) For the model in Equation (3.16), give the state variable S^n and the exogenous information variable W^n.

b) Give the stochastic gradient algorithm for this problem. Unlike the previous problem, this gradient will be a function of p^n.

The situation where the gradient depends on the price p^n is a fairly significant complication. What is happening here is that instead of trying to find an optimal solution x^* (or more precisely, $x^{\pi,N}$), we are trying to find a function $x^{\pi,N}(p)$.
The trick here is to pick a functional form for $x^{\pi,N}(p)$. We suggest two alternatives:

Lookup table- Even if p is continuous, we can discretize it into a series of discrete prices $p_1, \ldots, p_K$, where we pick the value p_k closest to a price p^n. Call this price p^n_k. Now think of a stochastic gradient algorithm indexed by whatever p_k is nearest to p^n. We then use the stochastic gradient to update $x^n(p^n_k)$ using

$$x^{n+1}(p^n_k) \quad = \quad x^n(p^n_k) + \alpha_n \nabla_x F^n(x^n, W^{n+1}).$$

Of course, we do not want to discretize p too finely. If we discretize prices into, say, 100 ranges, then this means we are trying to find 100 order quantities $x^{\pi,N}(p)$, which would be quite difficult.

Parametric model- Now imagine that we think we can represent the order quantity $x^{\pi,N}(p)$ as a parametric function

$$x^{\pi,N}(p|\theta) \quad = \quad \theta_0 + \theta_1 p + \theta_2 p^{\theta_3}. \tag{3.17}$$

When we use a parametric function such as this, we are no longer trying to find the order quantity $x^{\pi,N}$; instead, we are trying to find θ that determines the function (in this case, (3.17)). Our stochastic gradient algorithm now becomes

$$\begin{aligned}\theta^{n+1} &= \theta^n + \alpha_n \frac{dF^n(x^{\pi,N}(p^n|\theta^n), W^{n+1})}{d\theta}, \\ &= \theta^n + \alpha_n \frac{dF^n(x^{\pi,N}(p^n|\theta^n), W^{n+1})}{dx} \frac{dx^{\pi,N}(p^n|\theta^n)}{d\theta},\end{aligned}$$

Remember that θ^n is a four-element column vector, while x^n is a scalar. The first derivative is our original stochastic gradient

$$\frac{dF^n(x^{\pi,N}(p^n|\theta^n), W^{n+1})}{dx} = \begin{cases} p - c & x \leq W, \\ -c & x > W. \end{cases}$$

The second derivative is computed directly from the policy (3.17), which is given by

$$\begin{aligned}\frac{dx^{\pi,N}(p^n|\theta^n)}{d\theta} &= \begin{pmatrix} \frac{dx^{\pi,N}(p^n|\theta^n)}{d\theta_0} \\ \frac{dx^{\pi,N}(p^n|\theta^n)}{d\theta_1} \\ \frac{dx^{\pi,N}(p^n|\theta^n)}{d\theta_2} \\ \frac{dx^{\pi,N}(p^n|\theta^n)}{d\theta_3} \end{pmatrix} \\ &= \begin{pmatrix} 1 \\ p^n \\ (p^n)^{\theta_3} \\ \theta_2 (p^n)^{\theta_3} \ln p^n \end{pmatrix}.\end{aligned}$$

Using the parametric model can be very effective if the parametric form matches the true form of the function $x^{\pi,N}(p)$. The lookup table representation is more general, which can be a feature, but if the discretization is too fine, then it will require a much larger number of iterations to solve.

With these strategies in mind, consider the next three extensions:

3.10 — Return to the objective function in Equation (3.14) where the price is only revealed after we make the order decision, but now p^{n+1} depends on history, as in

$$p^{n+1} = p^n + \varepsilon^{n+1}.$$

Discuss how you might approach this problem, given what we presented above.

3.11 — Repeat Exercise 3.10, but now assume that

$$p^{n+1} = 0.5p^n + 0.5p^{n-1} + \varepsilon^{n+1}.$$

3.12 — Repeat Exercise 3.10, but now the quantity x^n is chosen subject to the constraint $0 \le x \le R^n$ where

$$R^{n+1} = \max\{0, R^n + x^n - W^{n+1}\},$$

and where the price $p = p^n$ is revealed before we make a decision. With this transition, our problem becomes a traditional inventory problem.

3.13 — A flexible spending account (FSA) is an accounting device that allows people to put away pre-tax money for the purpose of covering medical expenses. You have to allocate how much you want to have available in year $t+1$ at the end of year t. The challenge is that if you put too much into the account, you lose whatever is left over.
Let M_{t+1} be your medical expenses in year $t+1$, and let x_t be the amount you allocate at the end of year t to spend in year $t+1$. Let r be your marginal tax rate where $0 < r < 1$. Your total expenditures in year $t+1$ is given by

$$C(x_t, M_{t+1}) = x_t + \frac{1}{1-r}\max\{0, M_{t+1} - x_t\}. \tag{3.18}$$

You would like to use a stochastic gradient algorithm of the form

$$x_{t+1} = x_t + \alpha_t \bar{g}_{t+1}, \tag{3.19}$$

where

$$\bar{g}_{t+1} = (1-\eta)\bar{g}_t + \eta\frac{dC(x_t, M_{t+1})}{dx_t} \tag{3.20}$$

and where $0 < \eta < 1$ is a smoothing factor. For the stepsize, use

$$\alpha_t = \frac{\theta^{step}}{\theta^{step} + K_t - 1}, \tag{3.21}$$

where K_t counts how many times the derivative of the cost function has changed signs. That is

$$K_{t+1} = \begin{cases} K_t + 1 & \text{if } \frac{dC(x_t, M_{t+1})}{dx_t} \frac{dC(x_{t-1}, M_t)}{dx_{t-1}} < 0. \\ K_t & \text{otherwise.} \end{cases} \quad (3.22)$$

Your challenge is to decide on the stepsize parameter θ^{step} and the smoothing parameter η by formulating this problem as a sequential decision problem. Assume that you have access to a simulator to evaluate the performance of the stepsize rule.

We are going to begin by finding the optimal solution assuming that we know the distribution for M_{t+1}:

a) What is $\frac{dC(x_t, M_{t+1})}{dx_t}$? Remember that this is computed after M_{t+1} becomes known.

b) Find the optimal static solution by setting the derivative (from part (a)) equal to zero, and then solving for x^*. Assume that the cumulative distribution function $F^M(m) = Prob(M_{t+1} \leq m)$ is known.

Now we are going to model the sequential learning problem where we will not assume that the distribution of M_{t+1} is known:

c) What is the state variable for this dynamic system?

d) What is(are) the decision variable(s)?

e) What is the exogenous information?

f) What is the transition function? Remember that you need an equation for each element of the state variable.

g) What is the objective function? What are you optimizing over?

3.14 — We are going to assume that the price that we sell our gas changes from month to month. The monthly profit function would be given by

$$F_t(x_t, W_{t+1}) = p_{t+1} \min(x_t, W_{t+1}) - cx_t,$$

where D_{t+1} is the demand for electricity (in megawatt-hours) for month $t + 1$.

Assume for simplicity that the price process evolves as follows:

$$p_{t+1} = \begin{cases} p_t - 1 & \text{with probability } 0.2, \\ p_t & \text{with probability } 0.6, \\ p_t + 1 & \text{with probability } 0.1. \end{cases}$$

a) Rewrite the five elements of the model that you originally provided in Exercise 3.15, part (a). Note that instead of looking for x_t, you are now looking for $x_t(p_t)$. This means that instead of looking for a scalar, we are now looking for a function.

b) We are going to start by representing $x_t(p_t)$ as a lookup table function, which means that we are going to discretize p_t into a set of discrete prices $(0, 1, 2, \ldots, 50)$. Without doing any coding, describe the steps of the method you would use to estimate the function $x_t(p_t)$ (your description has to be careful enough that someone could write code from it). Compare the complexity of this problem to the basic model.

c) Repeat (b), but instead of a lookup table for $x_t(p_t)$, approximate the functional form for the policy using

$$x_t(p_t|\theta) = \theta_0 + \theta_1 p_t + \theta_2 \ln p_t + \theta_3 \exp\{\theta_4 p_t\}.$$

Again describe the steps of an adaptive algorithm to find θ.

Programming questions

These exercises use the Python module *AdaptiveMarketPlanning.*[1]

3.15 — A major industrial gas company, which converts air to liquified oxygen and nitrogen, has to sign contracts for natural gas for the production of electricity. The contracts provide a quantity of gas for the upcoming month, signed a month in advance. Let W_{t+1} be the demand for electricity (in mega-watt-hours) for month $t+1$, and let x_t be the quantity of gas, decided at the beginning of month t, to be purchased in month $t+1$ (we could have indexed this $x_{t,t+1}$).

Assume that we purchase gas (normally measured in units of millions of btus) at a price of \$20 per equivalent megawatt-hour (mwh), and sell

[1]See http://tinyurl.com/sdagithub

it at a price of \$26 per equivalent mwh (later we are going to introduce uncertainty into these prices).

For simplicity, we are going to assume that the random variables $W_1, W_2, \ldots, W_t$, are stationary, which means they all have the same distribution, but the distribution is unknown. Your profits for month t are given by

$$F_t(x_t, W_{t+1}) = p\min\{x_t, W_{t+1}\} - cx_t.$$

Further assume that you are going to use a stochastic gradient algorithm for finding the order quantities x_t, given by

$$x_{t+1} = x_t + \alpha_t \nabla F_t(x_t, W_{t+1}).$$

Finally, assume that the stepsize is given by

$$\alpha_t = \frac{\theta^{step}}{\theta^{step} + N_t - 1},$$

where N_t counts the number of times that the gradient changes signs. We write the updating equation for N_t using

$$N_{t+1} = \begin{cases} N_t + 1 & \text{if } \nabla F_{t-1}(x_{t-1}, W_t)\nabla F_t(x_t, W_{t+1}) < 0, \\ N_t & \text{otherwise.} \end{cases}$$

a) Write out the five elements of the model for this problem. For the objective function, you want to find the best policy (this will be an algorithm) to maximize total profits from buying and selling natural gas over a horizon of $T = 24$ months. Note that the search over policies refers to finding the best value of θ^{step}.

b) Use the python package *AdaptiveMarketPlanning*[2] to evaluate $\theta^{step} = (2, 5, 10, 20, 50)$ for the model in part (a).

c) How would your objective function change if you were to optimize the terminal reward rather than the cumulative reward? Be sure to write out the expected in its nested form (that is, using notation like $\mathbb{E}W$ if you are taking an expectation over W).

d) Repeat the search for the best θ^{step} (using the same values), but now using the final reward formulation that you gave in part c.

[2] See https://tinyurl.com/sdagithub

e) Now assume that your objective function is given by

$$F_t(x_t, W_{t+1}) = p_{t+1} \min(x_t, W_{t+1}) - cx_t.$$

where we now assume that we have to sign our contract for a quantity x_t without knowing the price we will receive on the electricity that we sell to the market. Instead, the price pt+1 is revealed during month $t + 1$. How would this change affect your model and solution strategy?

4

Learning the Best Diabetes Medication

4.1 Narrative

When people find they have high blood sugar, typically evaluated using a metric called "A1C," there are several dozen drugs that fall into four major groups:

- Sensitizers - These target liver, muscle, and fat cells to directly increase insulin sensitivity, but may cause fluid retention and therefore should not be used for patients with a history of kidney failure.
- Secretagoges - These drugs increase insulin sensitivity by targeting the pancreas but often causes hypoglycemia and weight gain.
- Alpha-glucosidase inhibitors - These slow the rate of starch metabolism in the intestine, but can cause digestive problems.
- Peptide analogs - These mimic natural hormones in the body that stimulate insulin production.

The most popular drug is a type of sensitizer called metformin, which is almost always the first medication that is prescribed for a new diabetic,

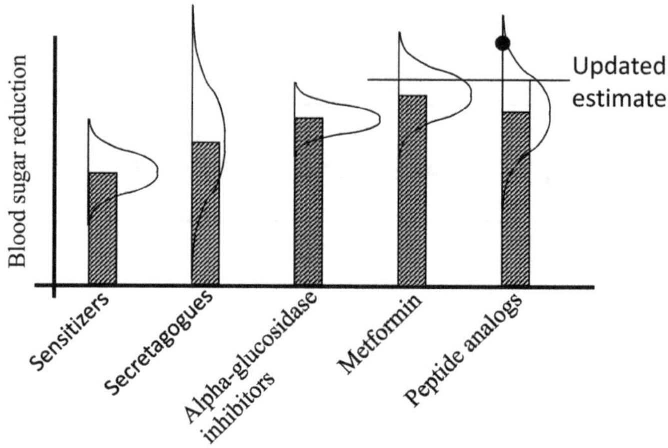

Figure 4.1: Beliefs about the potential that each drug might have on the reduction of blood sugar.

but this does not always work. Prior to working with a particular patient, a physician may have a belief about the potential of metformin, and drugs from each of the four groups, to reduce blood sugar that is illustrated in Figure 4.1.

A physician will typically start with metformin, but this only works for about 70 percent of patients. Often, patients simply cannot tolerate a medication (it may cause severe digestive problems). When this is the case, physicians have to begin experimenting with different drugs. This is a slow process, since it takes several weeks before it is possible to assess the effect a drug is having on a patient. After testing a drug on a patient for a period of time, we observe the reduction in the A1C level, and then use this observation to update our estimate of how well the drug works on the patient.

Our challenge is to find a policy for identifying the medication that achieves the greatest possible reduction in a patient's A1C level.

4.2 Basic model

For our basic model, we are going to assume that we have five choices of medications: metformin, or a drug (other than metformin) drawn

Table 4.1: Metformin and the four drug classes and the average reduction over the full population.

Drug	A1C reduction	Std. dev.
Metformin	0.32	0.12
Sensitizers	0.28	0.09
Secretagoges	0.30	0.17
Alpha-glucosidase inhibitors	0.26	0.15
Peptide analogs	0.21	0.11

from one of the four major drug groups. Let $\mathcal{X} = \{x_1, x_2, x_3, x_4, x_5\}$ be the five choices. From observing the performance of each drug over hundreds or thousands of patients, it is possible to construct a probability distribution of the reduction in A1C levels across all patients. The results of this analysis are shown in Table 4.1 which reports the average reduction and the standard deviation across all patients. We assume that the distribution of reductions in A1C across the population is normally distributed, with means and standard deviations as given in the table.

To create a model, let

$\bar{\mu}^0_x$ = the mean reduction in the A1C for drug choice x across the population,

$\bar{\sigma}^0_x$ = the standard deviation in the reduction in A1C for drug x.

Our interest is learning the best drug for a particular individual. Although we can describe the patient using a set of attributes, for now we are only going to assume that the characteristics of the patient do not change our belief about the performance of each drug for an individual patient.

We do not know the reduction we can expect from each drug, so we represent it as a random variable μ_x, where we assume that μ_x is normally distributed, which we write as

$$\mu_x \sim N(\bar{\mu}^0_x, (\bar{\sigma}^0_x)^2).$$

We refer to the normal distribution $N(\bar{\mu}^0_x, (\bar{\sigma}^0_x)^2)$ as the *prior distribution of belief* about μ_x.

We index each iteration of prescribing a medication by n, starting at 0, which refers to the time before we have run any experiments. Assume that we always observe a patient for a fixed period of time (say, a month). If we try a drug x on a patient, we make a noisy observation of the true value μ_x of the patient's response to a medication. Assume we make a choice of drug x^n using what we know after n trials, after which we observe the outcome of the $n+1^{st}$ trial, which we denote W^{n+1} (this is the reduction in the A1C level). This can be written

$$W^{n+1} = \mu_{x^n} + \varepsilon^{n+1}.$$

Remember that we do not know μ_x; this is a random variable, where $\bar{\mu}_x^n$ is our current estimate of the mean of μ_x.

4.2.1 State variables

Our state variable is our belief about the random variable μ_x which is the true effect of each drug on a particular patient after n trials. S^0 is the initial state, which we write as

$$S^0 = (\bar{\mu}_x^0, \bar{\sigma}_x^0)_{x \in \mathcal{X}},$$

where we also include in S^0 the assumption of normality, which remains through all the experiments. After n experiments, the state is

$$S^n = (\bar{\mu}_x^n, \bar{\sigma}_x^n)_{x \in \mathcal{X}},$$

where we no longer include the normality assumption because it is captured in our initial state (the distribution is static, so by convention we do not include it in the dynamic state variable).

Later, we are going to find it useful to work with the *precision* of our belief, which is given by

$$\beta_x^n = \frac{1}{(\bar{\sigma}_x^n)^2}.$$

We can then write our state variable as

$$S^n = (\bar{\mu}_x^n, \beta_x^n)_{x \in \mathcal{X}}.$$

We are using what is known as a *Bayesian belief model*. In this model, we treat the unknown value of a drug, μ_x, as a random variable with initial prior distribution given by S^0. After n experiments with different drugs, we obtain the *posterior* distribution of belief S^n.

4.2.2 Decision variables

The decision is the choice of medication to try for a month, which we write as

$$\begin{aligned} x^n &= \text{the choice of medication,} \\ &\in \mathcal{X} = \{x_1, \ldots, x_5\}. \end{aligned}$$

We are going to determine x^n using a policy $X^\pi(S^n)$ that depends only on the state variable S^n (along with the assumption of the normal distribution in S^0).

4.2.3 Exogenous information

After making the decision x^n, we observe

$$\begin{aligned} W_x^{n+1} &= \text{the reduction in the A1C level resulting from the drug} \\ & \quad x = x^n \text{ we prescribed for the } n+1^{st} \text{ trial.} \end{aligned}$$

A reader might wonder why we write the information learned from the decision x^n is written W_x^{n+1} rather than W_x^n. We do this to capture the information available in each variable. Thus, the decision x^0 depends only on the initial state S^0. The state S^n for $n \geq 1$ depends on S^0 along with the observations $W_{x^0}^1, \ldots, W_{x^{n-1}}^n$, but not $W_{x^n}^{n+1}$, since we have not yet made completed the $n+1^{st}$ experiment that would reveal $W_{x^n}^{n+1}$. By letting $W_{x^n}^{n+1}$ be the result of the prescription x^n, we know that x^n cannot depend on W^{n+1}, which would be like seeing into the future.

4.2.4 Transition function

The transition function captures how the observed reduction in A1C, W_x^{n+1}, affects our belief state S^n. Although it takes a little algebra, it is possible to show that if we try drug $x = x^n$ and observe W_x^{n+1}, we can update our estimate of the mean and precision using

$$\bar{\mu}_x^{n+1} = \frac{\beta_x^n \bar{\mu}_x^n + \beta^W W_x^{n+1}}{\beta_x^n + \beta^W}, \tag{4.1}$$

$$\beta_x^{n+1} = \beta_x^n + \beta^W, \tag{4.2}$$

where β^W is the precision of an observation (we can make this dependent on x if necessary). For all $x \neq x^n$, $\bar{\mu}^n_x$ and β^n_x remain unchanged.

The transition function, which we earlier wrote as a generic function

$$S^{n+1} = S^M(S^n, x^n, W^{n+1})$$

in Equation (1.16), is given by Equations (4.1) - (4.2).

4.2.5 Objective function

Each time we prescribe a drug $x = x^n$, we observe the reduction in the A1C represented by $W^{n+1}_{x^n}$. We want to find a policy that chooses a drug $x^n = X^\pi(S^n)$ that maximizes the expected total reduction in A1C. Our canonical model used $C(S^n, x^n, W^{n+1})$ as our performance metric. For this problem, this would be

$$C(S^n, x^n, W^{n+1}) = W^{n+1}_{x^n}.$$

We write the problem of finding the best policy as

$$\max_\pi \mathbb{E}\left\{\sum_{n=0}^{N-1} W^{n+1}_{x^n}|S_0\right\}, \tag{4.3}$$

where $x^n = X^\pi(S^n)$, and $S^{n+1} = S^M(S^n, x^n, W^{n+1})$. Here, the conditioning on S_0 is particularly important because it carries the prior distribution of belief.

4.3 Modeling uncertainty

Sampling random outcomes for our asset selling problem was relatively simple. For our medical setting, generating outcomes of the random variables $W^1, \ldots, W^n, \ldots$ is a bit more involved.

With the asset selling problem, we were generating random variables with mean 0 and a given variance which we assumed was known. In this medical application, the reduction in the A1C from a particular drug is a noisy observation of the true mean μ_x (for a particular patient) which we can write as

$$W^{n+1} = \mu_x + \varepsilon^{n+1},$$

where ε^{n+1} is normally distributed with mean 0 and a variance (which we assume is known) given by $(\sigma^W)^2$. The real issue is that we do not know μ_x. Given what we know after n experiments with different drugs, we assume that μ_x is normally distributed with mean $\bar{\mu}_x^n$ and precision β_x^n. We write this as

$$\mu_x|S^n \sim N(\bar{\mu}_x^n, \beta_x^n)$$

where we use the precision β_x^n instead of the more customary variance when we write our normal distribution. We then write the distribution of W^{n+1} as conditioned on μ_x using

$$W^{n+1}|\mu_x \sim N(\mu_x, \beta_x^W).$$

This means that we have to simulate two random variables: the true performance of drug x on our patient, given by μ_x (given our beliefs after n experiments), and then the noise ε^{n+1} when we try to observe μ_x. This just means that instead of generating one normally distributed random variable, as we did in our asset selling problem, we have to generate two.

4.4 Designing policies

A popular class of policies for this problem class is organized under the general name of *upper confidence bounding*. One of the first UCB policies has the form

$$X^{UCB}(S^n) = \arg\max_{x \in \mathcal{X}} \left(\bar{\mu}_x^n + 4\sigma^W \sqrt{\frac{\log n}{N_x^n}} \right), \tag{4.4}$$

where N_x^n is the number of times we have tried drug x (recall that "$\arg\max_x$" returns the value of x that achieves the maximum). It is standard practice to replace the coefficient $4\sigma^W$ by a tunable parameter, giving us

$$X^{UCB}(S^n|\theta^{UCB}) = \arg\max_{x \in \mathcal{X}} \left(\bar{\mu}_x^n + \theta^{UCB} \sqrt{\frac{\log n}{N_x^n}} \right). \tag{4.5}$$

A popular variant that we have found works surprisingly well was originally introduced under the name *interval estimation*, which his given by

$$X^{IE}(S^n|\theta^{IE}) \quad = \quad \arg\max_{x\in\mathcal{X}} \left(\bar{\mu}_x^n + \theta^{IE}\bar{\sigma}_x^n\right), \qquad (4.6)$$

where $\bar{\sigma}_x^n$ is the standard deviation of the estimate $\bar{\mu}_x^n$.

The policies (4.5)-(4.6) both share the structure of choosing the drug x that maximizes our estimate of its performance $\bar{\mu}_x^n$ plus a term that is often called an "uncertainty bonus." The intuition behind these policies is that the estimates $\bar{\mu}_x^n$ may be low because of bad luck. Without the uncertainty bonus, a few poor outcomes may mean that we may never try a drug again. These policies have attracted considerable attention from the research literature that can derive theoretical bounds on their performance, but ultimately it all depends on experimental comparisons using realistic data. An important step in the evaluation of the policies is the tuning of the parameter θ^{UCB} or θ^{IE}.

A third strategy that has attracted considerable attention is known as Thompson sampling. This approach takes a random sample from our belief about μ_x for each drug x, and then takes the best of these. More precisely, let

$$\hat{\mu}_x^n \sim N(\bar{\mu}_x^n, \theta^{TS}\bar{\sigma}_x^n)$$

be a random sample drawn from a normal distribution with mean $\bar{\mu}_x^n$ and standard deviation $\bar{\sigma}_x^n$, which is our current belief about the true response μ_x. The parameter θ^{TS} is a tunable parameter that influences the uncertainty we have around the estimated mean $\bar{\mu}_x^n$.

Now choose the drug to try next using

$$X^{TS}(S^n|\theta^{TS}) = \arg\max_{x\in\mathcal{X}} \hat{\mu}_x^n. \qquad (4.7)$$

Thompson sampling favors choices where the estimated performance $\bar{\mu}_x^n$, given what we know after n observations (across all drugs), but randomizes the performance. The randomization encourages exploration, since drugs whose estimated impact on A1C may not be the highest, still have a chance of coming out with the highest sampled value $\hat{\mu}_x^n$.

We note that all three of these policies, $X^{UCB}(S^n|\theta^{UCB})$, $X^{IE}(S^n|\theta^{IE})$, and $X^{TS}(S^n|\theta^{TS})$, all share two characteristics: the policy itself requires solving an optimization problem (the "$\arg\max_x$"), and they all have tunable parameters.

4.5 Policy evaluation

We originally wrote our objective function as

$$\max_\pi \mathbb{E}\left\{\sum_{n=0}^{N-1} W_{x^n}^{n+1}|S_0\right\},$$

but writing the expectation in this way is a bit vague. Recall that we have two sets of random variables: the true values of μ_x for all $x \in \mathcal{X}$, and the observations $W^1, \ldots, W^N$ (or more precisely, the noise when we try to observe μ_x). We can express this nested dependence by writing the objective function as

$$\max_\pi \mathbb{E}_\mu \mathbb{E}_{W^1,\ldots,W^N|\mu}\left\{\sum_{n=0}^{N-1} W_{x^n}^{n+1}|S_0\right\}. \tag{4.8}$$

There are two ways to simulate the value of a policy:

Nested sampling - First we simulate the value of the truth μ_x for all $x \in \mathcal{X}$ where we let $\psi \in \Psi$ be a sample realization of μ, which we write as $\mu(\psi)$. We then simulate the observations W, where we let $\omega \in \Omega$ be a sample realization of $W^1(\omega), \ldots, W^N(\omega)$, which means that ω is an outcome of all possible observations over all possible drugs $x \in \mathcal{X}$, over all experiments $n = 1, \ldots, N$.

Simultaneous sampling - Here, we let ω be a sample realization of both μ_x and the observations $W^1, \ldots, W^N$.

If we use nested sampling, assume that we generate K samples of the true values $\mu(\psi_k)$, and L samples of the errors $\varepsilon^1(\omega_\ell), \ldots, \varepsilon^N(\omega_\ell)$. For the sampled truth $\mu(\psi_k)$ and noise $\varepsilon^n(\omega_\ell)$, the performance of drug x^n in the $n+1^{st}$ experiment would be

$$W_{x^n}^{n+1}(\psi_k, \omega_\ell) \quad = \quad \mu(\psi_k) + \varepsilon^n(\omega_\ell).$$

We can then compute a simulated estimate of the expected performance of a policy using

$$\overline{F}^{\pi} = \frac{1}{K}\sum_{k=1}^{K}\left(\frac{1}{L}\sum_{\ell=1}^{L}\sum_{n=0}^{N-1} W_{x^n}^{n+1}(\psi_k,\omega_\ell)\right),$$

where $x^n = X^{\pi}(S^n)$ and

$$S^{n+1}(\psi_k,\omega_\ell) = S^M(S^n(\psi_k,\omega_\ell), X^{\pi}(S^n(\psi_k,\omega_\ell)), W^{n+1}(\psi_k,\omega_\ell)).$$

If we use simultaneous sampling, then a sample ω determines both the truth $\mu(\omega)$ and the noise $\varepsilon(\omega)$, allowing us to write a sampled estimate of our observation $W_{x^n}^{n+1}$ as

$$W_{x^n}^{n+1}(\omega_\ell) = \mu(\omega_\ell) + \varepsilon^n(\omega_\ell).$$

The estimated value of a policy is given given by

$$\overline{F}^{\pi} = \frac{1}{L}\sum_{\ell=1}^{L}\sum_{n=0}^{N-1} W_{x^n}^{n+1}(\omega_\ell).$$

If we use one of our parameterized policies where θ is the tunable parameter, we might write the expected performance as $\overline{F}^{\pi}(\theta)$. Then, the optimization problem would be

$$\max_{\theta} \overline{F}^{\pi}(\theta), \tag{4.9}$$

which we can solve using a variety of search procedures which we review in more depth in Chapter 7.

4.6 Extensions

We have been describing a problem that applies to a single patient. This means that we would have to solve this problem from scratch for each patient. If we have a million diabetes patients, then we would have a million models.

Imagine that we would like to use information from different patients to learn a single model. We can do this by characterizing each patient using a set of attributes $a = (a_1, \ldots, a_K)$. Assume for the moment that each element a_k is discrete (e.g. gender) or discretized (e.g. age, divided into ranges). In fact, we are going to start by assuming that there is a single attribute, gender. Let G^n the gender of the nth patient. Now we have two forms of exogenous information: the gender G^n of the nth patient, and the outcome W^n of the treatment of the nth patient.

We start with a knowledge state K^0 that is our vector $(\bar{\mu}^0, \beta^0)$ introduced earlier in the chapter. The first patient will have gender G^1, which means our state variable (that is, everything we know) after the first patient arrives is $S^1 = (K^0, G^1)$. We then make a decision x^1 regarding the treatment of patient 1, after which we observe an outcome W^1 describing how the treatment worked. We use this information to obtain an updated knowledge state K^1, after which the process repeats.

$$\begin{aligned}(K^0, G^1, S^1 = (K^0, G^1), x^1, W^1, K^1, G^2, S^2 = (K^1, G^2), \ldots,\\ K^{n-1}, G^n, S^n = (K^{n-1}, G^n), x^n, W^n, K^n, G^{n+1}, \ldots)\end{aligned}$$

We pause for a moment and note that our indexing is different from what we used in the basic model. In our basic model, the index n referred to visits by a patient. We make a decision x^n *after* the nth visit using what is known from the first n visits. We let W^{n+1} be the outcome of this treatment, incrementing n to $n+1$ to emphasize that x^n was computed without knowing W^{n+1}.

With our new model, however, n refers to a patient. It makes more sense to let G^n be the gender of the nth patient, at which point we make a decision for the nth patient, and let W^n be the outcome of the treatment for the nth patient. We do not increment n until we see the $n+1^{st}$ patient, at which point we see the gender of the $n+1^{st}$ patient.

4.7 What did we learn?

- We introduced the idea of a sequential decision problem that is a pure learning problem, where the state variable consists only of belief state variables.

- We saw an example of a problem where the uncertainty was in the true value of the performance of a choice such as the diabetes medication.

- We introduced an example of a cost function approximation policy that is a form of parameterized optimization problem, and illustrated this idea using three types of policies: upper confidence bounding (which is a general class of policies), interval estimation, and Thompson sampling.
- We note that each policy involves a tunable parameter and formulate the problem of tuning as its own optimization problem.
- We showed how to model the presence of exogenous information variables (such as the gender of the patient) as a fully sequential decision problem, known in the learning literature as a "contextual bandit problem" (the context is the gender). Instead of finding the best x, we are now looking for the best $x(G)$ as a function of gender (we could expand this with other attributes of patients).

4.8 Exercises

Review questions

4.1 — What is the fundamental difference from an algorithmic perspective between the diabetes problem we solved in this chapter, and the problem solved in Chapter 3?

4.2 — When we let $\bar{\mu}_x^n$ be the estimate of how well the drug performs on a patient after n trials, what is n measuring? Is it the number of times we have tried drug x?

4.3 — What is the state variable for this problem?

4.4 — In Section 4.4, we introduced an upper confidence bounding policy, an interval estimation policy, and a policy based on Thompson sampling. What characteristics did these policies have in common?

4.5 — Did our objective function optimize cumulative reward or final reward? Why did we use that version? What changes if we switch to the other objective function in terms of searching for a good policy?

Problem solving questions

4.6 — You are trying to determine the dosage of a diabetes medication that produces the greatest reduction in blood sugar. You are currently experimenting with three dosages that we designate by d_1, d_2 and d_3. Let μ_i true reduction in blood sugar produced by dosage d_i. After n experiments of different drugs, let $\bar{\mu}_i^n$ be the estimate of the reduction produced by dosage d_i. We wish to exploit the observation that our beliefs about μ_i are correlated. Let $\sigma_{ii'} = Cov(\mu_i, \mu_{i'})$ be the covariance in our belief about μ_i and $\mu_{i'}$.
Assume that after n tests of different dosages gives us the current vector of estimates

$$\bar{\mu}^n = \begin{bmatrix} 32 \\ 42 \\ 20 \end{bmatrix}.$$

Assume that the variance of a single experiment is 16 and that our covariance matrix Σ^n is given by

$$\Sigma^n = \begin{bmatrix} 8 & 4 & 2 \\ 4 & 8 & 4 \\ 2 & 4 & 8 \end{bmatrix}.$$

a) Write out the equations for finding the updated estimates $\bar{\mu}^{n+1}$ and covariance matrix Σ^{n+1} given an observation W^{n+1}.

b) Assume that we try dosage d_2 and obtain an observation $W^{n+1} = 50$. Compute the updated estimates $\bar{\mu}^{n+1}$ and covariance matrix Σ^{n+1}.

4.7 — Show how to adapt the policy presented earlier to our problem where gender is the only patient attribute by using a lookup table representation, which means that instead of learning $\bar{\mu}_x^n$, we learn $\bar{\mu}_{a,x}^n$ where a = gender. So, instead of learning an estimate $\bar{\mu}_x^n$ for each treatment x, we have to learn an estimate $\bar{\mu}_{a,x}^n$ for each combination of gender $a = G^n$ and treatment $x = x^n$.

4.8 — Sketch a strategy for applying the ideas of this chapter to the market planning problem in Chapter 3.

4.9 — Is it possible to apply the methods of Chapter 3 to the diabetes problem? Explain.

4.10 — Now imagine that instead of just gender, we capture age by decade $(0-9, 10-19, ..., 80^{+})$, smoker or not, and race (assume eight categories of ethnicity), giving us an attribute vector $a = (a_{gender}, a_{age}, a_{smoker}, a_{race})$. If $a \in \mathcal{A}$, how many elements does $\mathcal{A}$ have? How would this impact your proposed solution in Exercise 4.7?

4.11 — Imagine that each element a_k in the attribute vector a has L possible values, and that a has K elements, which means that $\mathcal{A}$ has L^K elements. If $L = 10$, what is the largest value of K so that learning our attribute-based model is easier than learning a model for each of 7 million diabetes patients.

4.12 — Now imagine that our attribute space $\mathcal{A}$ is simply too large to be practical. What we have done up to now is a lookup table representation where we find an estimate $\bar{\mu}^n_{a,x}$, which becomes problematic when the number of possible values of a becomes large. An alternative approach is to use a parametric model. The simplest would be a linear model where we would write

$$\bar{\mu}_{a,x} = \sum_{f \in \mathcal{F}} \theta_f \phi_f(a, x),$$

where $\phi_f(a, x)$ for $f \in \mathcal{F}$ is a set of features that we (as analysts) would have to define. For example, one feature might just be an indicator of gender, or age range, or race. In this case, there would be a feature for each possible gender, each possible age range, and so on.

a) If there are L possible values of each K attributes, what is the minimum number of features we would need?

b) Suggest more complex features other than just those that indicate the value of each attribute.

c) Contrast the strengths and weaknesses of a lookup table representation versus our linear model.

4.13 — We are going to evaluate different policies for finding the best drug for reducing blood sugar. We assume that our prior distribution of belief for each medication is given in Table 4.1.

We begin with a learning policy known as interval estimation given by

$$X^{IE}(S^n|\theta^{IE}) = \arg\max_{x \in \mathcal{X}}(\bar{\mu}_x^n + \theta^{IE}\bar{\sigma}_x^n).$$

We use a Bayesian belief model where it is convenient to use the concept of *precision* which is simply one over the variance. So, the precision in our initial estimate of the true value μ_x is given by

$$\beta_x^0 = \frac{1}{(\sigma_x^0)^2},$$

where σ_x^0 is given in Table 4.1.

After n experiments, we are going to use our policy to make a decision x^n which is the drug to try for the n+1st experiment. We do not know the true performance μ_x of drug x, but we can observe it using a noisy observation of the true value μ_x which we write using

$$W_x^{n+1} = \mu_x + \varepsilon_x^{n+1}.$$

Assume that the standard deviation of a single experiment is $\sigma^W = 5$. We use the observation of W_x^{n+1} to update our beliefs using:

i) If we try drug x:

$$\bar{\mu}_x^{n+1} = \frac{\beta_x^n \bar{\mu}_x^n + \beta^W W_x^{n+1}}{\beta_x^n + \beta^W}, \tag{4.10}$$

$$\beta_x^{n+1} = \beta_x^n + \beta^W. \tag{4.11}$$

ii) If x is a drug we do not try, then:

$$\bar{\mu}_x^{n+1} = \bar{\mu}_x^n, \tag{4.12}$$

$$\beta_x^{n+1} = \beta_x^n. \tag{4.13}$$

Answer the following:

a) Using a Bayesian belief model, what is the state variable?

b) What is the transition function for the belief model?

c) Write out the expected value of a policy $X^\pi(S^n)$ using the expectation operator $\mathbb{E}$. Be sure to index the operator to indicate which random variables are involved, as in $\mathbb{E}_\mu$ or $\mathbb{E}_W$ (or $\mathbb{E}_{W_{1,\ldots,M}}$). You can show conditioning by using $\mathbb{E}_{W|\mu}$ (this is the expectation over the observed reduction W given we know the true mean μ).

4.14 — We might reasonably think that the parameter θ^{IE} should depend on the number of experiments remaining in our budget, which means that θ^{IE} needs to be a function of n (or equivalently, it would be a function of the remaining experiments $N - n$). There are two ways to represent this function. Discuss (without any programming) the strengths of each approach, and the computational challenges that would be involved.

a) Lookup table - Instead of searching over a scalar θ^{IE}, we would have to search over a vector θ^{IE}_n.

b) Parametric - We might assume a function form such as $\theta^{IE} = \theta^{slope}(N - n)$, where now we just have to tune the scalar θ^{slope}.

4.15 — We have approached this problem as if we are solving the problem for each patient. Imagine that we have I patients indexed by $i = 1, \ldots, I$, remembering that I might be 10 million patients. Finding a vector of estimates $\bar{\mu} = (\bar{\mu}_x)_{x \in \mathcal{X}}$ for each patient would be written $\bar{\mu} = (\bar{\mu}_i)_{i=1}^I$ where each $\bar{\mu}_i = (\bar{\mu}_{ix})_{x \in \mathcal{X}}$. Creating 10 million estimates seems a bit clumsy.

Imagine instead that each patient has a vector of attributes $a = (a_1, \ldots, a_M)$ where $a \in \mathcal{A}$. There may be a lot of attributes, in which case the set $\mathcal{A}$ would be quite large, but we may choose a small subset so that $\mathcal{A}$ is not as big, such as gender and whether they smoke. We can again use two different representations of $\bar{\mu}_{ax}$. As before, discuss the strengths and computational challenges of each of the following ways of modeling $\bar{\mu}_{ax}$:

a) Lookup table - We would enumerate each of the attributes $a \in \mathcal{A}$, and create an estimate $\bar{\mu}_{ax}$ of the performance of each drug x and for each attribute a. This might be a large set, but should be smaller than 10 million.

b) Parametric - This requires coming up with a parametric form for $\bar{\mu}_{ax}$ for each drug x. One might be

$$\bar{\mu}_{ax} = \sum_{f \in \mathcal{F}} \bar{\theta}_{fx} \phi_f(a).$$

The functions $\phi_f(a)$ are sometimes called basis functions (other terms are independent variables or covariates). These might be indicator variables that capture, for example, the gender of the patient or whether they are a smoker. This representation replaces computing $\bar{\mu}_{ax}$ for each attribute a with computing a vector of coefficients $\bar{\mu}_{ax}$ for a set of features. The set $\mathcal{F}$ is presumably much smaller than the set of attributes (if this is not the case, then we should use the lookup table representation).

Programming questions

These exercises use the Python module *AdaptiveMarketPlanning*.[1]

4.16 — Perform $L = 1000$ simulations of the interval estimation policy over a budget of $N = 20$ experiments using $\theta^{IE} = 1$. Let $\hat{F}^{IE}$ be the performance of the IE policy for a particular sample path. Make the assumption that the true performance of a drug, μ_x, is given in Table 4.2, and use the assumptions for the standard deviation of each belief from Table 4.1. Also use the standard deviation $\sigma^W = 5$ for the experimental variation as we did in Exercise 4.13.

a) Compute the mean and standard deviation of the value of the policy $\overline{F}^{IE}(\theta^{IE})$ with $\theta^{IE} = 1$.

b) Evaluate the IE policy for $\theta^{IE} = (0, 0.2, 0.4, \ldots, 2.0)$ and plot $\overline{F}^{IE}(\theta)$. What do you learn from this plot?

[1]See http://tinyurl.com/sdagithub

Table 4.2: True values for a particular patient

Drug	A1C reduction	Truth
Metformin	0.32	0.25
Sensitizers	0.28	0.30
Secretagoges	0.30	0.28
Alpha-glucosidase inhibitors	0.26	0.34
Peptide analogs	0.21	0.24

4.17 — Evaluate the IE policy given a budget $N = 20$ over the values $\theta^{IE} = (0, 0.2, 0.4, \ldots, 2.0)$ for two different sets of truths:

a) First assume that the prior is $\mu_x^0 = 0.3$ for all drugs x and where the initial standard deviation $\sigma_x^0 = 0.10$. This means we are assuming that the truth $\mu_x \sim N(\bar{\mu}_x^0, (\bar{\sigma}_x^0)^2)$. However, we are going to sample our truth using

$$\hat{\mu}_x = .3 + \varepsilon$$

where ε is uniformly distributed in the interval $[-0.15, +0.15]$. This is an example of having a prior distribution of belief (in this case, that is normally distributed around 0.3) but sampling the truth from a different distribution (that is uniformly distributed around the mean 0.3).

Perform 10,000 repetitions of each value of θ^{IE} to compute the average performance. What conclusions can you draw from the resulting plot over the 11 values of θ^{IE}?

b) For this exercise we are going to simulate our truth from the prior using

$$\mu_x = \bar{\mu}_x^0 + \varepsilon$$

where $\bar{\mu}^0$ is given in Table 4.2 ("A1C reduction") and where ε is uniformly distributed in the interval $[-.5\bar{\mu}_x^0, +.5\bar{\mu}_x^0]$. Perform 10,000 repetitions of each value of θ^{IE} to compute the average performance. What conclusions can you draw from the plot?

5

Stochastic Shortest Path Problems - Static

Shortest path problems over graphs are both an important application area (arising in transportation, logistics, and communication), but are also a fundamental problem class that arises in many other settings. The most familiar shortest path problem is the classical deterministic problem illustrated in Figure 5.1, where we have to find the best path from node 1 to node 11, where the cost of traversing each arc is known in advance.

Shortest path problems build on a fundamental dynamic programming recursion. Let

$\mathcal{N}$ = the set of all nodes in the network (the nodes $1, 2, \ldots, 11$),

$\mathcal{N}_i^+$ = the set of all nodes that can be reached directly from node i,

$\mathcal{N}_j^-$ = the set of all nodes that are connected to node j,

$\mathcal{L}$ = set of all links (i, j) in the network,

c_{ij} = the cost of traversing link (i, j), where j is assumed to be in the set $\mathcal{N}_i^+$.

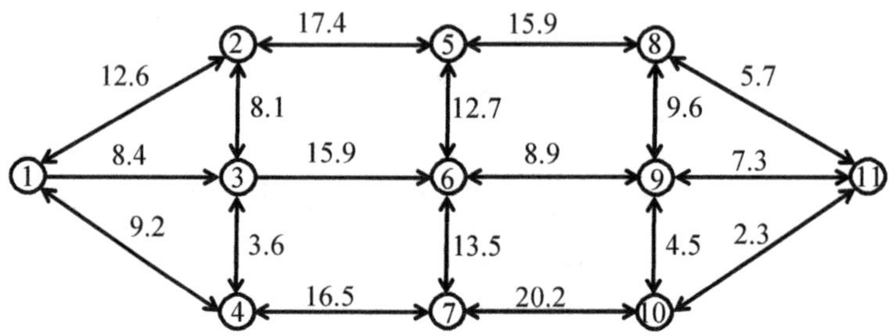

Figure 5.1: Network for a deterministic shortest path problem.

Let v_i be the minimum cost from node i to the destination node 11. The values v_i for all nodes $i \in \mathcal{N}$ should satisfy

$$v_i = \min_{j \in \mathcal{N}_i^+} (c_{ij} + v_j). \tag{5.1}$$

We can execute Equation (5.1) by initializing v_{11} to zero, and setting all other values to some large number. If we loop over every node i and compute v_i using Equation (5.1) repeatedly, the values v_i will converge to the optimal value. This is a very inefficient version of a shortest path algorithm.

Another way to view our network is to assume that each node i is a state S, and let $V_t(S_t)$ be the value of being in state S_t at "time" t. In our shortest path problem, we are going to use t to index the number of links we have traversed on our way from node 1 to the node represented by S_t.

From a state (node) S_t, assume we make a decision we call "x" which would be a decision to traverse a link emanating from the node corresponding to state S_t. We can write this set of decisions as $\mathcal{X}_s$ representing the decisions x that are available to use when we are in state $S_t = s$.

Next let $C(s,x)$ be the cost of being in state s and choosing decision x, which would correspond to our link cost c_{ij} in our network above. Finally, we are going to use a "state transition function" that we denote by $S^M(s,x)$ that tells us what state we transition to if we are in state s and take action $x \in \mathcal{X}_s$.

Using this notation, we can rewrite Equation (5.1) as

$$V_t(s) = \min_{x \in \mathcal{X}_s} \left(C(s,x) + V_{t+1}(S_{t+1}) \right). \tag{5.2}$$

where $S_{t+1} = S^M(s,x)$. We can execute Equation (5.2) by setting $V_T(s) = 0$ for a large enough value of T (that is, the largest number of links that we might traverse in a path). Since we might set T too large, we have to add to the set of choices in $\mathcal{X}_s$ the ability to stay at the destination node at time T. We then set $t = T - 1$ and run (5.2) for all states s. We keep repeating this until we get to $t = 0$. When we run the system this way, the time index t is really a counter for how many links we have traversed.

Equation (5.2) is a deterministic version of what is known as Bellman's equation. In the remainder of this exercise, we are going to show how to use Bellman's equation to handle uncertainty in our shortest path problem.

5.1 Narrative

You are trying to create a navigation system that will guide a driverless vehicle to a destination over a congested network. We assume that our system has access to both historical and real-time link costs, from which we can create estimates of the mean and variance of the cost of traversing a link. We can think of this as a shortest path problem where we see distributions rather than actual costs, as depicted in Figure 5.2.

We are going to start by assuming that we have to make decisions of which link to traverse based on these distributions. After we traverse a link from i to j, we then experience a sample realization from the distribution. We want to choose a path that minimizes the expected costs.

5.2 Basic model

We are going to assume that we are trying to traverse the network in Figure 5.2 starting at a node q and ending at a destination r.

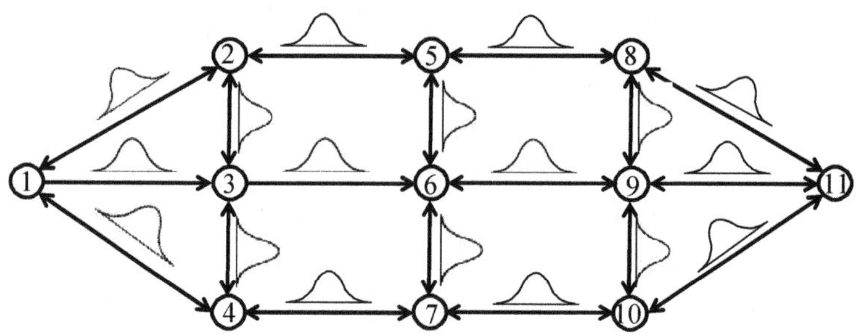

Figure 5.2: Network for a stochastic shortest path problem where distributions are known, but costs are not observed until after decisions are made.

5.2.1 State variables

In this basic problem, the state $S_t = N_t$ is the node where we are located after t link traversals. For the basic problem, it is more natural to say that we are in "state" (node) i, but the state variable will become a lot more interesting in the extensions.

5.2.2 Decision variables

We are modeling the decision as the node j that we traverse to given that we are at node i. There is a large community that works on problems that fit this class where the decision is represented as an action a, where a takes on one of a set of discrete values in the set $\mathcal{A}_s$ when we are in state s.

A convenient way to represent decisions is to define

$$x_{tij} = \begin{cases} 1 & \text{if we traverse link } i \text{ to } j \text{ when we are at } i, \\ 0 & \text{otherwise.} \end{cases}$$

This notation will prove useful when we write our objective function.

5.2.3 Exogenous information

After traversing the link (i, j), we observe

$\hat{c}_{tij}$ = the cost we experience traversing from i to j during the tth traversal (which we only observe after traversing the link).

For the moment, we are going to assume that the new observation $\hat{c}_{tij}$ is stored in a very large database. We can then use these to estimate the average cost $\bar{c}_{ij}$ of traversing link (i,j) (we exclude the index t for $\bar{c}_{ij}$ since this is the average cost regardless of when we traverse link (i,j)).

5.2.4 Transition function

For our basic graph problem, if we make decision $x_{tij} = 1$, the state $N_t = i$ evolves to state $N_{t+1} = j$.

5.2.5 Objective function

We can model our costs using the following notation:

$\hat{c}_{tij}$ = A random variable giving the cost to traverse from node i to node j.

$\bar{c}_{ij}$ = An estimate of the expected value of $\hat{c}_{tij}$ computed by averaging over our database of past travel costs.

$\bar{\sigma}_{ij}$ = Our estimate of the standard deviation of $\bar{c}_{ij}$ computed using historical data.

We assume that we have to make the decision of which link to traverse out of a node i before seeing the actual value of the random cost $\hat{c}_{tij}$. This means that we have to make our decision using our best estimate of $\hat{c}_{tij}$, which would be $\bar{c}_{ij}$.

We could write our objective function using

$$\min_{x_{tij},(i,j)\in\mathcal{L}} \sum_{t=0}^{T} \sum_{i\in\mathcal{N}} \sum_{j\in\mathcal{N}_i^+} \hat{c}_{tij} x_{tij}, \tag{5.3}$$

but this formulation would require that we know the realizations $\hat{c}_{tij}$. Instead, we are going to use the expectation, giving us

$$\min_{x_{tij},(i,j)\in\mathcal{L}} \sum_{t=0}^{T} \sum_{i\in\mathcal{N}} \sum_{j\in\mathcal{N}_i^+} \bar{c}_{ij} x_{tij}. \tag{5.4}$$

The optimal solution of this problem would be to set all $x_{tij} = 0$, which means we do not get a path. For this reason, we have to introduce *constraints* of the form

$$\sum_{j \in \mathcal{N}_q^+} x_{tqj} = 1, \tag{5.5}$$

$$\sum_{i \in \mathcal{N}_r^-} x_{t-1,ir} = 1, \tag{5.6}$$

$$\sum_{i \in \mathcal{N}_j^-} x_{t-1,ij} - \sum_{k \in \mathcal{N}_j^+} x_{tjk} = 0, \quad \text{for } j \neq q, r, \tag{5.7}$$

$$x_{tij} \geq 0, \quad (i,j) \in \mathcal{L}\ 0 \leq t \leq T. \tag{5.8}$$

Equations (5.4) - (5.8) represent a *linear program*, and there are powerful packages that can be used to solve this problem when written this way. However, there are specialized algorithms (known simply as "shortest path algorithms") that take advantage of the structure of the problem to produce exceptionally fast solutions.

However, this approach does not provide a method for handling uncertainty. Below, we describe how to solve the stochastic shortest path problem using our language of designing policies, which will provide a foundation for addressing uncertainty.

5.3 Modeling uncertainty

For our basic model we are only using the point estimates $\bar{c}_{ij}$ which we assume is just an average of prior observations collected using, for example, travel cost estimates drawn from GPS-enabled smartphones. When estimates are based on field observations, the method is called *data-driven*, which means we do not need a model of uncertainty - we just need to observe it.

For example, let $\bar{c}_{ij}$ be our current estimate of the average travel cost for link (i, j) and assume we just observed a cost of $\hat{c}_{tij}$. We might update our estimate using

$$\bar{c}_{ij} \leftarrow (1 - \alpha)\bar{c}_{ij} + \alpha\hat{c}_{tij},$$

where α is a smoothing parameter (sometimes called a learning rate or a stepsize) that is less than 1.

If we update our estimates in this way, then it means that the vector of estimated travel times $\bar{c}$ is varying dynamically, although we might only do updates once each day, as opposed to within a trip. In our modeling framework, the vector of cost estimates $\bar{c}$ are captured by the initial state S_0. If we let n index the day of the trip, we would let $\bar{c}^n$ be the cost estimates using the first n days of data, which is then held in the initial state S_0^n when planning for day $n+1$.

5.4 Designing policies

Our "policy" for this deterministic problem is a function that maps the "state" (that is, what node we are at) to an action (which link we move over). We can solve this problem by optimizing the linear program represented by Equations (5.4) - (5.8), which gives us the vector x_{ij}^* for all links (i,j). We can think of this as a function where given the state (node i) we choose an action, which is the link (i,j) for which $x_{ij} = 1$. We can write this policy as a function $X^\pi(S_t)$ using

$$X^\pi(S_t = N_t = i) = j \quad \text{if } x_{ij} = 1.$$

Alternatively, we can solve Bellman's equation as we did initially for our deterministic shortest path problem using Equation (5.1). This gives us a value v_i which is the minimum travel cost from each node i to the destination node r. Once these values are computed, we can make decisions using the following policy

$$X^\pi(i) = \arg\min_{j \in \mathcal{N}_i^+}(\bar{c}_{ij} + v_j). \tag{5.9}$$

This means that our "stochastic" shortest path problem can be solved just as we solved our deterministic problem. Below in our extensions, we show that with a minor twist, the situation changes dramatically.

5.5 Policy evaluation

Policy evaluation for this problem is not required, because the policy is optimal. Although our link costs are stochastic, as long as we do not learn anything about the actual cost until after we make our decision, the

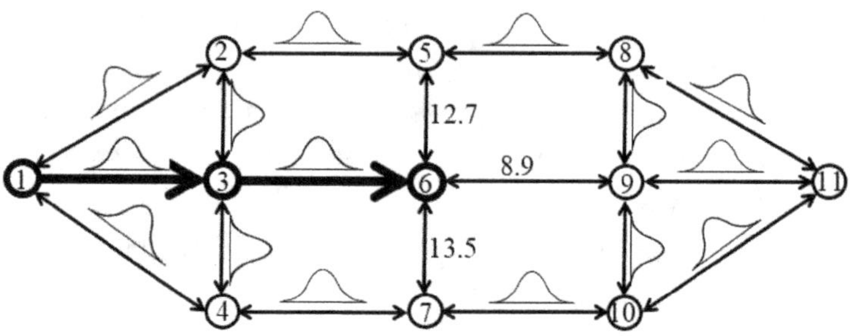

Figure 5.3: Network for a stochastic shortest path problem where travelers get to see the link costs before making a decision. This graph depicts a traveler who has traversed the path 1-3-6, and now sees the costs on links out of node 6.

optimal decisions involve solving a deterministic shortest path problem. This will be the last time in this book that we see a problem like this.

Under extensions, we are going to introduce uncertainty in a way that allows us to introduce a powerful algorithmic strategy called *approximate dynamic programming* (also known as *reinforcement learning*).

5.6 Extension - Adaptive stochastic shortest paths

We are going to change the information that we can use while making decisions. In our first stochastic shortest path problem, we assumed that we had to choose the next link to traverse *before* we see the actual travel cost over the link. Now assume that we make our decision *after* we observe the link costs, which means we make our decision using the actual cost $\hat{c}_{ij}$ rather than its expectation (or average) $\bar{c}_{ij}$. This is illustrated in Figure 5.3, where a traveler at node 6 gets to see the actual costs on the links out of node 6 (rather than just knowledge of the distributions).

If we pretend for the moment that someone can give us the values v_j which is the minimum travel cost from node j to our destination node r, an optimal policy for choosing the next downstream node would be written

$$X^\pi(i) = \arg\min_{j \in \mathcal{N}_i^+} (\hat{c}_{ij} + v_j). \tag{5.10}$$

The problem here is that we cannot compute v_j as we did before using Equation (5.1). In fact, the switch to seeing the costs before we make a decision requires a fundamental change to our basic model.

In our deterministic model, or static stochastic model, the state variable S_t after "t" transitions was the node i where the traveler was located. This was the only information we need at that point in time.

In our new stochastic model, just capturing the node where our traveler is located is no longer enough. Remember that above, we introduced a state variable as "all the information we need at time t from history to model the system from time t onward." Later, in Chapter 7, we are going to provide a more precise definition, but for now this will serve our needs.

Our new stochastic shortest path problem introduces new information that is needed to make a decision: the costs out of the node where we are located. We are going to find it convenient to introduce two types of state variables:

N_t = The physical state of the system, which is usually controlled directly by decisions.

I_t = Other information that we need to make a decision.

In our network problem, our physical state would be the node where we are located, while the "other information" variable I_t would capture the costs on links out of the node where we are located, which we write as

$$I_t = (\hat{c}_{tij}), i = N_t, j \in \mathcal{N}_i^+.$$

Assume that at time t that $N_t = i$. We are going to add time t to our index for link costs, which means we will replace $\hat{c}_{ij}$ with $\hat{c}_{tij}$ to refer to the cost when we go from i to j at time t. We might then write

$$\begin{aligned} S_t &= (N_t, I_t) \\ &= (i, (\hat{c}_{tij})_{j \in \mathcal{N}_i^+}). \end{aligned}$$

To see what this does to our previous way of solving our shortest path problem, take a fresh look at Bellman's equation as we first introduced it in Equation (5.2) which becomes

$$V_t(S_t) = \min_{x_t \in \mathcal{X}_s} \left(C(S_t, x_t) + V_{t+1}(S_{t+1}) \right), \tag{5.11}$$

where S_{t+1} would be given by

$$S_{t+1} = (N_{t+1}, I_{t+1}),$$

where:

$$N_{t+1} = \text{The node produced by our decision } x \text{, so if } x_{ij} = 1, \text{ then } N_{t+1} = j.$$

$$I_{t+1} = \text{The costs that are observed out of node } N_{t+1} \text{, which depends on the decision } x_t. \text{ If } x_t \text{ sends us to node } j \text{ so that } N_{t+1} = j, \text{ then } I_{t+1} = (\hat{c}_{t+1,jk_1}, \hat{c}_{t+1,jk_2}, \hat{c}_{t+1,jk_3}) \text{ (assuming there are three links out of node } j).$$

Assume that $N_t = i$. The cost function $C(S_t, x)$ is given by

$$C(S_t, x) = \sum_{j \in \mathcal{N}_i^+} \hat{c}_{tij} x_{ij}.$$

Remember that S_t (where $N_t = i$) contains the costs $\hat{c}_{tij}$ for the links (i, j) out of i, so these are known (and contained in S_t).

Equation (5.11) is easy to write, but hard to solve. We are going to first describe two computational challenges. Then, we are going to introduce the idea of the post-decision state to solve one of the two challenges. Finally, we are going to provide a brief introduction to a class of methods known as approximate dynamic programming (but often called reinforcement learning) to handle the second challenge.

5.6.1 Computational challenges

We start by identifying two computational challenges:

- The link costs I_{t+1} out of downstream node N_{t+1} (which is determined by the decision x) are not known. Said differently, I_{t+1} is a random variable at time t, which means we cannot even calculate $V_{t+1}(S_{t+1})$. We fix this by taking the expectation, which we write as

$$V_t(S_t) = \min_{x \in \mathcal{X}_s} \left(C(S_t, x) + \mathbb{E}\{V_{t+1}(S_{t+1}) | S_t, x\} \right). \tag{5.12}$$

  The expectation operator $\mathbb{E}$ should be viewed as taking an average over the possible link costs that a traveler might encounter once they arrive at node N_{t+1} when making decision x (which determines N_{t+1}).

To write this more explicitly, assume that x_t sends us to node j (which means that $x_{tij} = 1$), and when we arrive we see $I_{t+1} = (\hat{c}_{t+1,jk_1}, \hat{c}_{t+1,jk_2}, \hat{c}_{t+1,jk_3})$. We learn these costs when we arrive to node j at time $t+1$, but they are random when we are at node i at time t thinking about what to do.

- The state space - Even if we assume that the costs $\hat{c}_{t+1,j}$ are discrete, the state space just grew dramatically. Imagine that we have discretized costs into buckets of 20 values. If there are three links out of every node, our state space has grown from the number of nodes, to one that is $20 \times 20 \times 20 = 8,000$ times larger.

To illustrate the challenge of computing the expectation, assume that each cost $\hat{c}_{t+1,jk}$ can take on values $c_1, c_2, \ldots, c_L$ with probabilities $p_{jk}(c_\ell)$. For example, c_1 might be 1 minute, c_2 might be 2 minutes, and so on. The probability $p_{jk}(c_\ell)$ is the probability that $\hat{c}_{t+1,jk} = c_\ell$.

Now assume that the decision x takes us to node j, after which we face a choice of traveling over links (j,k_1), (j,k_2) or (j,k_3). We would compute our expectation using

$$\begin{aligned}\mathbb{E}\{V_{t+1}(S_{t+1})|S_t, x\} &= \sum_{\ell_1=1}^{L} p_{jk_1}(c_{\ell_1}) \sum_{\ell_2=1}^{L} p_{jk_2}(c_{\ell_2}) \sum_{\ell_3=1}^{L} p_{jk_3}(c_{\ell_3}) \\ &\quad \times V_{t+1}(S_{t+1} = (j, (c_{\ell_1}, c_{\ell_2}, c_{\ell_3}))). \qquad (5.13)\end{aligned}$$

To put it bluntly, Equation (5.13) is pretty ugly. Those triple summations are going to be hard to compute.

Compounding the problem is the size of the state space. To use Bellman's equation in Equation (5.12) (or (5.2)), we have to compute $V_t(S_t)$ for every possible state S_t. When the state was just a node, that is not too bad, even if there are thousands (even tens of thousands) of nodes. However, adding the information variable I_t to the state makes the problem dramatically harder.

To see how quickly this grows the state space, imagine that there are 20 possible values of each cost variable $\hat{c}_{tij}$. That means there are 8,000 possible values of I_t. If our network has 10,000 nodes (that is, N_t can take on 10,000 values), then S_t can now take on $10,000 \times 8,000 = 80,000,000$ values.

This is our first peek at what happens when a state variable becomes a vector. The number of possible values of the state variable grows exponentially, a process known widely as the *curse of dimensionality.*

5.6.2 Using the post-decision state

All is not lost for this problem. There is a trick we can use that allows us to overcome the curse of dimensionality for this particular problem. The main computational challenge with Bellman's equation in Equation (5.12) is the expectation operator, which is easily the most dangerous piece of notation solving sequential decision problems.

We are going to use two powerful strategies to overcome this problem in this setting (and we are going to use these strategies for other settings). First, we introduce the idea of the *post-decision state* which we designate S_t^x. The post-decision state is the state of the system immediately *after* we make a decision, and before any new information arrives, which is why we index it by t.

To see pre- and post-decision states, return to Figure 5.3. As we saw before, our pre-decision state (which we call "the state") S_t is

$$S_t = (6, (12.7, 8.9, 13.5)).$$

Once we have made a decision, we are still at node 6, but imagine that the decision we made was to go to node 9. We might describe our physical post-decision state $R_t^x = 9$, which we might alternatively state as "going to node 9." However, we no longer need those troublesome observations of costs on the links out of node 6, given by $(\hat{c}_{t+1,6,5}, \hat{c}_{t+1,6,9}, \hat{c}_{t+1,5,7}) = (12.7, 8.9, 13.5)$. This means that our post-decision state is

$$S_t^x = (9).$$

Using the post-decision state, we are going to break Bellman's equation into two steps. Instead of going from S_t to S_{t+1} to S_{t+2} as we are doing in Bellman's Equation (5.12), we are going to first step from the pre-decision state S_t to post-decision state S_t^x which we do by rewriting Equation (5.12) as

$$V_t(S_t) = \min_{x \in \mathcal{X}_s} \left(C(S_t, x) + V_t^x(S_t^x)\right), \tag{5.14}$$

where V_t^x is the value of being at post-decision state S_t^x. Note that we no longer have the expectation, because by construction the post-decision state involves a given decision x_t (e.g. "go to node 9") but no new information (which is the random part). Note that in this case, the post-decision state S_t^x consists of just the node, which means it is much simpler than S_t.

We are not out of the woods. We still have to compute $V_t^x(S_t^x)$, which is done using

$$V_t^x(S_t^x) \quad = \quad \mathbb{E}\{V_{t+1}(S_{t+1})|S_t, x\}. \tag{5.15}$$

So, we still have to compute that expectation, and it has not gotten any easier. Assume that our decision x is to go to node j (which means that $x_{ij} = 1$), and let

$$\hat{c}_{t+1,j} = (\hat{c}_{t+1,jk}, \ k \in \mathcal{N}_j^+)$$

be the set of link costs out of node j. Our next pre-decision state S_{t+1} would then be

$$S_{t+1} \quad = \quad (j, \hat{c}_{t+1,j}).$$

Now assume that we have a way of sampling possible values of $\hat{c}_{t+1,j}$. We might do this from a database of historical observations of link costs, or we might build a probability distribution from past data and sample from this. Assume we are going to do this iteratively, and let $\hat{c}_{t+1,ij}^n$ be the nth sample of the link cost from i to j. We can use this sampling-based strategy to create an estimate of the expectation, rather than the exact value. This is done in the next section.

5.6.3 Approximate dynamic programming

Using post-decision state variables solves the problem of computing the expectation while finding the best decision x_t, but we still have the issue of dealing with the large state space. For this, we are going to turn to the methods that are broadly known as *approximate dynamic programming*, where we replace the post-decision value function $V_t^x(S_t^x)$ with an approximation.

We are going to construct approximations $\overline{V}_t^x(j)$ of the value of being at node j, where

$$\overline{V}_t^{x,n}(S_t^x = j) \quad \approx \quad \mathbb{E}\{V_{t+1}(S_{t+1})|S_t^x\}.$$

Let $\overline{V}_t^{x,n}(j)$ be our approximation of $\mathbb{E}\{V_{t+1}(S_{t+1})|S_t^x\}$ after observing n samples. One way to build this approximation is to use samples of the value of being at node j. Imagine that we are going to pass forward through the network, making decisions using approximations $\overline{V}_t^{x,n-1}(S_t^x)$ obtained from previous iterations, along with sampled costs $\hat{c}_{tij}^n$. We can obtain a sampled estimate of the value of being in state S_t using

$$\hat{v}_t^{x,n}(i) \quad = \quad \min_{j \in \mathcal{N}_i^+} \left(\hat{c}_{tij}^n + \overline{V}_t^{x,n-1}(S_t^x = j)\right). \tag{5.16}$$

We are then going to use $\hat{v}_t^{x,n}(i)$, which is the value of being in state S_t (which includes both the node i and the costs $\hat{c}_{tij}^n$ for all j out of node i), to update the previous post-decision state S_{t-1}^x, which we do using

$$\overline{V}_{t-1}^{x,n}(i) = (1-\alpha_n)\overline{V}_{t-1}^{x,n-1}(i) + \alpha_n \hat{v}_t^{x,n}(i). \tag{5.17}$$

Here, α_n is known as a smoothing factor or learning rate, but for technical reasons it is also known as a "stepsize." We might use a constant such as $\alpha_n = .1$ or $.05$, but a common strategy is to use a declining formula such as

$$\alpha_n = \frac{\theta^\alpha}{\theta^\alpha + n - 1}, \tag{5.18}$$

where θ^α is a tunable parameter. For example, if we set $\theta^\alpha = 1$, we get $\alpha_n = 1/n$. In this case, it is possible to verify the Equation (5.17) is averaging over the values $\hat{v}_t^n(i)$. In practice, this formula is unlikely to work well for this problem because the stepsize goes to zero too quickly.

We pause to note two advantages of the use of the post-decision state S_t^x:

- We no longer have to deal with the expectation when optimizing over the choice of output links from a node (see Equation (5.16)).

- Approximating the value function $\overline{V}_t^{x,n}(S_t^x = i)$ is much simpler since the post-decision state S_t^x is now just a scalar, which is much easier to estimate than a higher-dimensional function.

One challenge with Equation (5.16) is that we are going to need initial values for $\overline{V}_t^{x,0}(i)$. A natural choice would be to solve the deterministic version of this problem where the costs $\hat{c}_{tij}$ are set equal to estimates of their means, and then obtain initial estimates of the cost incurred to get from each node to the destination.

An alternative method is to use the estimates $\overline{V}^{x,n-1}(i)$ to make decisions using

$$x_t^n(i) \quad = \quad \arg\min_{j \in \mathcal{N}_i^+} \left(\hat{c}_{tij}^n + \overline{V}_t^{x,n-1}(j)\right). \tag{5.19}$$

The decision $i_t^n = x_t^n(i)$ gives us the next node after node i based on the sampled costs $\hat{c}_{tij}^n$ and the estimates of the cost $\overline{V}_t^{x,n-1}(j)$ to get from node j to the destination node r. When we arrive at r, we have an entire path consisting of the nodes

$$(q, i_1^n, i_2^n, \ldots, r).$$

We also have the sampled costs $\hat{c}_{t,i_t^n,i_{t+1}^n}^n$ over the entire path. Assume there are T links in the path. We then traverse backward over the path starting with $\hat{v}_T^n(r) = 0$, and computing

$$\hat{v}_t^n(i_t^n) \quad = \quad \hat{c}_{t,i_t^n,i_{t+1}^n}^n + \hat{v}_{t+1}^n(i_{t+1}^n). \tag{5.20}$$

We then use these estimates in our smoothing process in Equation (5.17).

This procedure is a form of *approximate dynamic programming* (alternatively known as *reinforcement learning*). More specifically, it is a form of *forward approximate dynamic programming* since it progresses by stepping forward through time. We have illustrated a pure forward pass procedure using Equation (5.16), which requires single passes through the network, and a double pass procedure using Equation (5.20), which consists of first stepping forward through the graph simulating decisions, and then backward for updating the value of being at each state.

This method is very robust with respect to complex pre-decision states. For example, we do not care how many links might be going out of each node, but we are leveraging the fact that our post-decision state variable is fairly simple (in this case, it is just the node where we are sitting).

5.7 What did we learn?

- This is the first (and only) time we have a problem where we can find the optimal policy for a sequential decision problem. Although we are optimizing over a stochastic network, the traveler does not receive any advance information about a link before traveling over the link, which means he has to make his choice based on expected costs.
- Since the basic model reduces to a deterministic shortest path problem, we can solve this optimally, which is a rare example of being able to solve the base problem optimally (in fact, this is the only time this will happen in this book).
- We then introduce the dimension that costs are revealed before the traveler traverses the link. We reformulate the problem, showing that the state variable now becomes much more complex, consisting of the node where the traveler is located and the costs of links out of the node. This problem can no longer be solved exactly using dynamic programming.
- We introduce and describe an approximate dynamic programming algorithm using the concept of a post-decision state variable which eliminates the expectation imbedded in Bellman's equation, and reduces the state space back to just the set of nodes.
- This is an example of a VFA policy. Since we cannot compute the value functions exactly, we cannot guarantee that it is an optimal policy.

5.8 Exercises

Review questions

5.1 — In the original stochastic shortest path problem in Section 5.2 where we only observe the actual cost after we traverse the link, explain why this can be solved exactly as a simple deterministic shortest path problem.

5.2 — For the version where we observe the actual cost over a link before we choose which direction to move, give the pre-decision and post-decision state variables.

5.3 — In Equation (5.17), we use the sampled value $\hat{v}_t^{x,n}(i)$ of being in state S_t to update the estimated value of being at the previous post-decision state given by $\overline{V}_{t-1}^{x,n}(i)$. Create a small numerical example to illustrate this equation.

Problem solving questions

5.4 — A traveler needs to traverse the graph shown in Figure 5.4 from node 1 to node 11. There is a probability that each link can be traversed, which is shown on the graph (these probabilities are known in advance). When the traveler arrives at nodes i, she gets to see which links (if any) can be traversed out of node i. If no links can be traversed, then the trip stops with failure. The goal is to choose a path that maximizes the product of these probabilities, but she is limited to traveling over links that are available.

a) Describe an appropriate state variable for this problem (with notation).

b) Imagine the traveler is at node 6 by following the path 1-2-6 and then sees that links 6-9 and 6-10 are available (but 6-8 is not available); what is her (pre-decision) state? I am looking for the numerical values of the state variables you provided in part (a).

c) Assume that the traveler is able to move to node 9, and decides to do this. What is the post-decision state after making this decision?

d) Write out Bellman's equation that characterizes the value of being in the pre-decision state after traversing the path 1-2-6 in terms of the downstream pre-decision states. Numerically calculate the value of being in the state after traversing 1-2-6 and observing that 6-9 and 6-10 are available (but 6-8) is not.

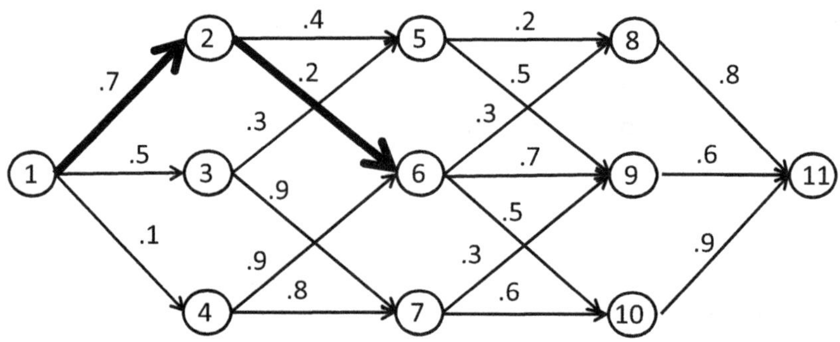

Figure 5.4: A shortest path problem to maximize the probability of completing a path.

5.5 — (This question applies to the stochastic shortest path problem described in Section 5.6). Write out the steps involved in fitting a value function approximation for the post-decision states by answering:

a) Write out the pre- and post-decision states. If the network has N nodes, and if the costs on each link are discretized into 20 values (assume at most L links out of any node), what is the size of the pre- and post-decision state spaces?

b) Give the equation for computing $\hat{v}_t^n(i)$. Is this the estimate of being at a pre-decision state or a post-decision state? Explain.

c) What does the "time" index t refer to?

d) Give the updating equation for updating the value at being at a post-decision state.

e) Why do we need the value of being at a post-decision state rather than a pre-decision state?

5.6 — Figure 5.5 illustrates the choices that an Uber driver might face. At node 1, she has a choice of trips (2-4) and (3-5). Assume that trips are at least 15 minutes, and trips are expected to be served within 10 minutes or they are lost. This means the trips out of nodes 6, 7 and 8 only become known after trips (2-4) and (3-5) would have been completed.

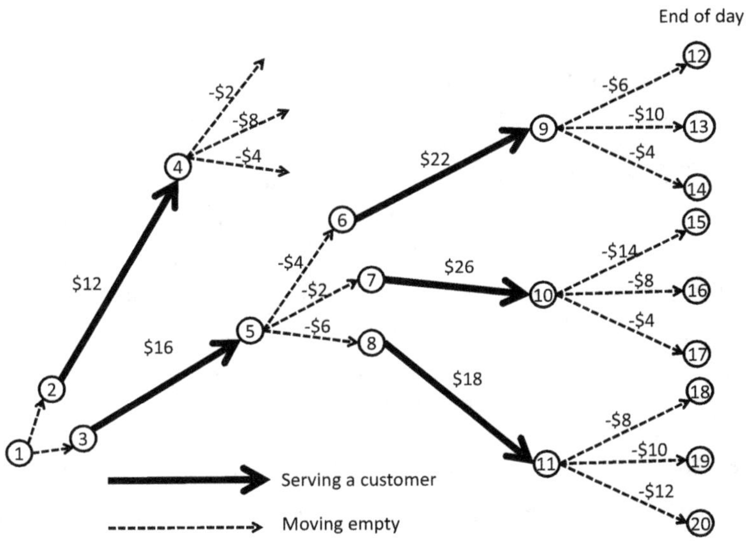

Figure 5.5: Possible choices for a single Uber driver.

Assume that our driver first chooses (3-5) and then chooses (7-10). After completing (7-10), she has to move to a recharging station to charge her battery before returning home (assume she chooses the least cost recharging station).
Alongside each movement is how much she makes (denote this by c_{ij}), which is positive for serving a customer and negative for moving empty. Of course she is trying to maximize profits over her entire shift.
The state of our driver is her location (node number) or the node number to which she is headed, along with any other available information known at that time relevant to her decision.

a) Given that she is initially at node 1, what is her (pre-decision) state? What is her post-decision state after deciding to accept trip (3-5)?

b) Let s_1 be the (pre-decision) state after serving trip (3-5). Let $\hat{v}_1(s_1)$ be the value of being in state s_1. What is $\hat{v}_1(s_1)$?

c) Let s_0^x be the previous post-decision state before s_1, and let $\hat{v}_0^x(s_0^x)$ be the value of being in s_0^x. What is $\hat{v}_0^x(s_0^x)$?

d) What is the computational advantage of using post-decision states over pre-decision states in terms of computing a policy? This should be a one-sentence response.

Programming questions

These exercises use the Python module *StochasticShortestPath_Static.*[1]

5.7 — (This question applies to the stochastic shortest path problem described in Section 5.6). Currently the algorithm has a fixed smoothing factor for estimating the value function approximations when solving the modified problem in the extension. Implement a declining stepsize, as follows:

$$\alpha_n = \frac{\theta^{step}}{\theta^{step} + n - 1}.$$

Run the python module for $\theta^{step} = (1, 5, 10, 20, 50)$ for 100 iterations, and compare the performance in terms of both the rate of convergence and the final solution. Which would you choose?

[1]See http://tinyurl.com/sdagithub

6

Stochastic Shortest Path Problems - Dynamic

6.1 Narrative

We are going to tackle stochastic shortest paths again, but this time we are going to do it just as it is done in Google maps (or any commercial navigation system). We all recognize that transportation networks often have predictable patterns of congestion, along with random variations that happen in the natural course of events. For example, an accident might create a backlog where we might estimate how the travel delays might evolve as a result of the accident.

The point of departure from the static shortest path problem is that our estimates of costs in the future are evolving over time. We are going to return to the problem where costs are stochastic, but when we arrive at a node i, we do not see the actual realizations of the costs out of node i. However, we are going to assume that we are given updated estimates of costs over the entire network. These estimates can be viewed as a forecast; we are going to assume that the actual cost that we incur when we traverse an arc will, on average, equal the forecast (that is, the forecasts are unbiased), but these forecasts will evolve over time as we get updates on the status of the network.

6.2 Basic model

Assume that when we have to make a decision at time t, we have an updated estimate of travel costs based on *current* congestion levels (by tracking the speed at which our smartphones are moving through traffic). We are going to represent these times using

$$\bar{c}_{tk\ell} = \text{the estimated cost of traversing link } (k,\ell) \text{ at time } t, \text{ using estimates based on what we know at time } t.$$

For now, we are not going to try to model the cost if we arrive at a point in time $t' > t$ given what we know at time t. So, we may estimate, at 3pm, that we are going to arrive at a link at 5pm, but we are going to use our 3pm estimate (as Google does now).

6.2.1 State variables

A traveler at node $N_t = i$ at time t is assumed to be given a set of forecasts

$$\begin{aligned} \bar{c}_t &= (\bar{c}_{ttk\ell})_{k,\ell\in\mathcal{N}}, \\ &= \text{the vector of estimates of the cost to traverse link } (k,\ell) \text{ at time } t, \text{ given what is known at time } t. \end{aligned}$$

The traveler's state S_t at time t is then

$$S_t = (N_t, \bar{c}_t).$$

Note that this state variable is *very* large; it consists of a vector of estimates of link costs for *every* link in the network.

6.2.2 Decision variables

The decision variables are the same as with the static stochastic shortest path problem

$$x_{tij} = \begin{cases} 1 & \text{if we traverse link } i \text{ to } j \text{ when we are at } i \text{ at time } t, \\ 0 & \text{otherwise.} \end{cases}$$

This decision has to obey the constraint that we do *something* when we are in state $N_t = i$ as long as i is not the destination. We write this constraint as

$$\sum_j x_{t,i,j} \quad = \quad 1 \text{ for } N_t = i \text{ other than the destination.} \tag{6.1}$$

If we are at the destination, then we do nothing, and instead write $x_{tij} = 0$ for i equal to the destination, and j any other node.

As above, we let $X^\pi(S_t)$ be our policy for determining the vector x_t which we assume has to satisfy the constraint (6.1).

6.2.3 Exogenous information

There are two types of exogenous information for this problem. The first type is the observed costs

$\hat{c}_{t+1,ij}$ = the actual cost of traversing link (i,j) after the traveler made the decision at time t to traverse this link.

Note that we only observe $\hat{c}_{t+1,ij}$ if the traveler traverses link (i,j) (we can just insert 0 for links we do not traverse, since we will not use these values).

The second type of new information is the updates to the estimates $\bar{c}_t$ of the link costs. We are going to model the exogenous information as the change in the estimates:

$$\begin{aligned} \delta\bar{c}_{t+1,k\ell} &= \begin{cases} \bar{c}_{t+1,k\ell} - \bar{c}_{tk\ell} & \text{if } x_{tk\ell} = 1, \\ 0 & \text{otherwise.} \end{cases} \\ \delta\bar{c}_{t+1} &= (\delta\bar{c}_{t+1,k\ell})_{(k,\ell)\in\mathcal{N}}. \end{aligned}$$

Our exogenous information variable, then, is given by

$$W_{t+1} \quad = \quad (\hat{c}_{t+1}, \delta\bar{c}_{t+1}).$$

6.2.4 Transition function

We are assuming that $\hat{c}_{t+1}$ arrives as exogenous information (we could have let the exogenous information be the change in the costs, but this is more natural).

The transition function for the forecasts evolves according to

$$\bar{c}_{t+1,k\ell} = \bar{c}_{tk\ell} + \delta\bar{c}_{t+1,k\ell}. \tag{6.2}$$

Finally, we update the physical state N_t using

$$N_{t+1} \quad = \quad \{j | x_{t,N_t,j} = 1\}. \tag{6.3}$$

In other words, if we are at node $i = N_t$ and we make the decision $x_{tij} = 1$ (which requires that we be at node i, since otherwise $x_{tij} = 0$), then $N_{t+1} = j$.

The updating of $\hat{c}_{t+1}$, Equation (6.2) for the forecasts $\bar{c}_{t+1}$ and Equation (6.3) for our physical state R_t, make up our transition function

$$S_{t+1} \quad = \quad S^M(S_t, X^\pi(S_t), W_{t+1}).$$

6.2.5 Objective function

We now write our objective function as

$$\min_\pi \mathbb{E}\left\{\sum_{t=0}^{T} \sum_{(i,j)\in\mathcal{N}} \hat{c}_{t+1,i,j} X^\pi(S_t)|S_0\right\}. \tag{6.4}$$

Note that our policy $X^\pi(S_t)$ makes the choice of the next link we move to given what we know at time t, captured by S_t.

6.3 Modeling uncertainty

In practice, the dynamic updating of costs (and forecasts) come from real systems, which means they are *data driven.* When this is the case, we do not use a mathematical model of the link costs. The alternative is to have a mathematical model of the random information W^{n+1}.

If we wish to run simulations, then we face the challenge of modeling the realization of the costs captured by $\hat{c}_t$, as well as the sequence of forecasts. Considerable care has to be given to this model. First, the change in the estimate of $\bar{c}_t$, which we represent by $\delta\bar{c}_{t+1}$, has to be drawn from a distribution with mean 0. In addition, the realizations $\hat{c}_{t+1}$ have to be drawn from a distribution with mean $\bar{c}_t$.

We do not want to minimize the challenge of creating a realistic stochastic model. Changes in link costs arise from different sources, from natural traffic variations, weather, accidents, and shifts in flows due to drivers responding to congestion elsewhere in the network. Stochastic variations in link costs are nonstationary, and are not independent, either over time or across links. However, beyond recognizing the difficult challenges, a more realistic model is beyond the scope of our discussion.

6.4 Designing policies

A quick hint that we are not going to be using Bellman's equation (even approximately) is the size of the state variable, which now includes forecasts of travel costs on every link in the network.

Instead, we are going to base our policy on a special model that we call a *lookahead model.* For example, at time t we can create a model consisting of states S_t, decisions x_t and exogenous information W_{t+1}, but in our base problem the state variable $S_t = (N_t, \bar{c}_t)$, which is quite complicated.

Instead, we are going to create a simpler model where we first create a new set of variables that are typically approximations of the variables in the base model. We differentiate a new set of variables for our lookahead models by putting tilde's on the variables in the base model, and index them by two time indices: time t, which is the time at which a decision is being made, and a second index t' which is the time within the lookahead model.

The sequence of states, decisions and exogenous information in the lookahead model would then be written

$$(\tilde{S}_{tt}, \tilde{x}_{tt}, \widetilde{W}_{t,t+1}, \ldots, \tilde{S}_{tt'}, \tilde{x}_{tt'}, \widetilde{W}_{t,t'+1}, \ldots).$$

Our vector of costs $\bar{c}_t$ would then be replaced with the vector $\tilde{c}_{tt'}$. We now face the challenge of designing a lookahead policy that we could call $\tilde{X}_{tt'}(\tilde{S}_{tt'})$ that determines $\tilde{x}_{tt'}$ within the lookahead model. Below we propose two strategies, both of which can be solved using a simple shortest path algorithm.

6.4.1 A deterministic lookahead policy

We approximate the problem by assuming that the costs in the lookahead model, $\tilde{c}_{tt'}$, are fixed and equal to the current estimates which means we let

$$\tilde{c}_{tt'k\ell} = \bar{c}_{tk\ell}.$$

This means that we no longer have the exogenous information variables $\widetilde{W}_{tt'}$, which gives us a deterministic lookahead model.

This allows us to solve our lookahead model deterministically, treating the cost estimates $\tilde{c}_{tt',k\ell}$ as the correct cost rather than random variables. In this case, our state variable is once again simply the node where the traveler is located (within the lookahead model).

We can solve this problem with a standard shortest path algorithm which, as we saw in Chapter 5, is a deterministic dynamic program that we can solve with Bellman's equation, which we do by first finding the "value" of being at node i at time t' in our lookahead model. We can compute these values by setting the values at the end of our lookahead model for time t equal to zero

$$\widetilde{V}_{t,t+H}(i) = 0, \text{ for all } i.$$

Then, we step back in time (in the lookahead model) for $t' = t+H-1, t+H-2, \ldots, t$ and compute, for each node i:

$$\widetilde{V}_{tt'}(i) \quad = \quad \min_{j \in \mathcal{N}_i^+} (\tilde{c}_{tt',ij} + \widetilde{V}_{t,t'+1}(j)). \tag{6.5}$$

Our lookahead policy is then given by

$$\tilde{X}^{\pi}_{tt'}(\tilde{S}_{tt'} = i) = \arg\min_{j \in \mathcal{N}_i^+} (\tilde{c}_{tt',ij} + \widetilde{V}_{t,t'+1}(\tilde{S}_{t,t'+1} = j)). \tag{6.6}$$

Finally, the policy that we are going to implement in the base model, if we are at node i, would be

$$X^{\pi}_t(S_t = i) = \tilde{X}^{\pi}_{tt}(S_t = i).$$

This is the policy that we are using when we follow a navigation system. Making decisions based on a deterministic lookahead model is one of the

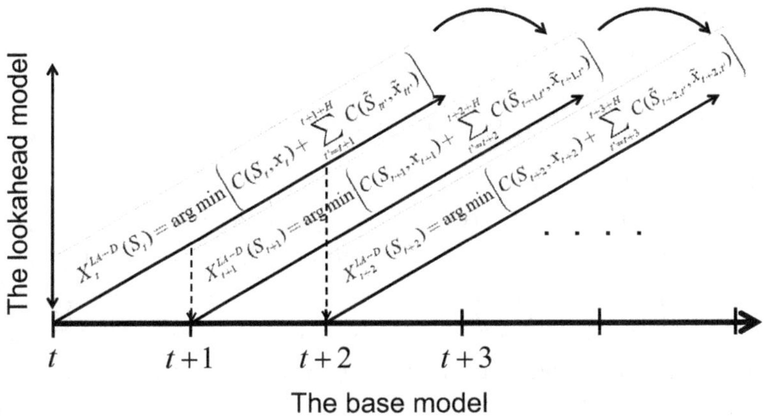

Figure 6.1: Illustration of simulating a direct lookahead policy, using a deterministic model of the future.

most widely used methods for making decisions in sequential decision problems under uncertainty.

Figure 6.1 illustrates a rolling lookahead process. At times t, $t+1$, $t+2$, …, we create and solve a lookahead model using estimates of costs as we know them. We then solve our shortest path problem, which is represented in the decisions $\tilde{x}_{tt'}(j)$ for all nodes j, but then we only implement the decision $\tilde{x}_{tt}(i)$ for the node i where we are located at time t.

When we hold a dynamically changing variable constant in a lookahead model, we refer to this variable as a *latent variable* in the lookahead model. This is one of a number of different types of approximations that can be made in a lookahead model. The most obvious approximation we are making is that we use a deterministic future which means that the estimates of link costs are being held constant within the lookahead model, even while they are changing in the base model.

The lookahead model, then, is its own model with its own characteristics, which is the reason why we use variables with tilde's - this is how we make the distinction between our base model, which uses variables such as S_t and x_t, and the lookahead model, where we use variables such as $\tilde{S}_{tt'}$ and $\tilde{x}_{tt'}$.

Next, we are going to propose a minor tweak to make this approach work better under uncertainty.

6.4.2 A parameterized deterministic lookahead policy

A simple strategy for handling uncertainty in our dynamic shortest path problem would be to replace our point estimate $\tilde{c}_{tt'k\ell} = \bar{c}_{tk\ell}$ for the cost of traversing link (k, ℓ) at time t with, say, the θ-percentile of the costs, suggesting that we write the costs as $\tilde{c}_{tt',k\ell}(\theta) = \bar{c}_{tij}(\theta)$. This logic could, for example, avoid a path through an area that sometimes becomes very congested, where the cost *might* be quite high.

This policy still produces a deterministic shortest path problem which is as easy to solve as when we used the point estimates $\bar{c}_t$. We simply modify Equations (6.5) - (6.6) above by using the θ-percentile link costs. We then designate the value functions $\tilde{V}_{tt'}(i|\theta)$ to indicate the dependence on the parameter θ, which is computed using

$$\tilde{V}_{tt'}(i|\theta) \quad = \quad \min_{j \in \mathcal{N}_i^+} \left(\tilde{c}_{tt',ij}(\theta) + \tilde{V}_{t,t'+1}(j|\theta)\right). \tag{6.7}$$

Our lookahead policy is then given by

$$\tilde{X}^{\pi}_{tt'}(\tilde{S}_{tt'} = i|\theta) = \arg\min_{j \in \mathcal{N}_i^+} \left(\tilde{c}_{tt',ij}(\theta) + \tilde{V}_{tt'}(\tilde{S}_{t,t'+1} = j)\right). \tag{6.8}$$

We then write our parameterized policy for the base model (which gives the decisions that are actually implemented) using

$$X^{\pi}_t(S_t = i|\theta) = \tilde{X}_{tt}(S_t = i|\theta).$$

This is equivalent to our original deterministic lookahead model, with one major exception: we need to tune θ by optimizing

$$\min_{\theta} \mathbb{E} \left\{ \sum_{t=0}^{T} C(S_t, X^{\pi}(S_t|\theta))|S_0 \right\}. \tag{6.9}$$

where $S_{t+1} = S^M(S_t, X^{\pi}(S_t|\theta), W_{t+1})$ (see Equations (6.2)-(6.3)) using some method for generating random realizations of $W_1, \ldots, W_T$.

The optimization problem in (6.9) is itself a challenging problem, but it is helped because θ is a scalar between 0 and 1. Practical algorithms for optimizing the objective function in (6.9) typically involve running simulations to get noisy observations of the function.

An obvious question is whether the use of a percentile different than $\theta = 0.5$ would improve the results. Our experience is that this is true when there is a penalty for late arrivals (for example, we want to arrive for an appointment at 9am).

6.5 What did we learn?

- We showed how to model a dynamic network problem, where the estimates of costs evolve over time. This time, the state of the system is the location of the traveler along with the estimates of costs on every link in the network.
- We introduced the idea of an approximate lookahead model, in this case a deterministic lookahead, which can be solved as a shortest path problem. Although this is an optimal solution, solving an approximate lookahead model, even optimally, is not an optimal policy.
- We describe latent variables, which are dynamic variables (the costs on the links) that are held constant in the lookahead model (which is why they are no longer in the state variable).
- We show how we can modify our deterministic lookahead into a parameterized deterministic lookahead. Instead of using the expected cost on each link, we might use the θ-percentile so that we consider how bad a link *might* be. The parameter θ has to be tuned, making this a hybrid of a deterministic lookahead approximation (a DLA policy) that is parameterized, which makes it a form of CFA policy, giving us a hybrid DLA/CFA policy.

6.6 Exercises

Review questions

6.1 — Why couldn't we use the approximate dynamic programming methods of Chapter 5 to solve our dynamic problem?

6.2 — How are we modeling the exogenous process W_t, in the lookahead model?

6.3 — Describe in words what we mean by a lookahead policy.

Problem solving questions

6.4 — We solve a deterministic lookahead model as our policy. This solves the deterministic problem optimally. Why isn't this an optimal policy?

6.5 — We solve our deterministic lookahead model (possibly with modified costs $\bar{c}_{ij}(\theta)$) as a deterministic dynamic program using Bellman's equation. Why wouldn't we then say that we are solving our base model using dynamic programming?

6.6 — Imagine that we want to leave as late as possible from the origin node, but there is a high penalty for arriving late at the destination. If we optimize over the θ-percentile of the costs $\bar{c}_{tij}(\theta)$, how might this logic help us avoid late arrivals?

6.7 — Given the insights from Exercise 6.6, how do you think using the θ-percentile costs would help for a problem where we are simply trying to minimize total travel time without regard to the possibility of arriving late?

6.8 — Provide the full model (state variables, decision variables, ...) for the setting where costs $\hat{c}_{tij}$ on links out of node i are revealed when the traveler arrives at node i and before she makes a decision which link to traverse. Remember that you are minimizing cumulative costs over the path. You do not have to design a policy; follow our standard practice of introducing a policy $X^\pi(S_t)$ without specifying the policy.

6.9 — Imagine that we want to solve our shortest path problem where we want to depart as late as possible from the origin, but we need to arrive to the destination by 9am. We assess a penalty η for each minute we arrive after $9am$. Describe a base model and parameterized lookahead policy for solving this problem. What is the state variable for the base model? What is the state variable for the lookahead model?

Programming questions

These exercises use the Python module *StochasticShortestPath_Dynamic*.[1]

6.10 — We are going to use a deterministic lookahead model as was done in the notes, but instead of using the expected cost on each link, we are going to use a percentile that we designate by θ^{cost}. For example, if $\theta^{cost} = 0.8$, then we would use the 80th percentile of the cost (think of this as using an estimate of how large the cost it might be). Let $\bar{c}_{tij}(\theta^{cost})$ be the θ^{cost}-percentile cost of link (i,j) given what we know at time t.

a) Write out the lookahead model, which would be a deterministic shortest path using costs $\bar{c}_{tij}(\theta^{cost})$ (as is done in the chapter). Use this model to formally define a lookahead policy $X^{DLA}(S_{tj}|\theta^{cost})$.

b) What is the state variable for the dynamic problem? Remember that the state variable includes all dynamically varying information used to make a decision (which includes computing costs and constraints), as well as computing the transition from t to $t+1$.

c) Write out the objective function used to evaluate our lookahead policy.

d) We now have a policy $X^{DLA}(S_{tj}|\theta^{cost})$ parameterized by θ^{cost}. Using the python module *StochasticShortestPath_Dynamic*, simulate the policy for $\theta^{cost} = (0, 0.1, 0.2, 0.3, 0.4, 0.5, 0.6, 0.7, 0.8, 0.9, 1.0)$. Simulate each version of the policy 100 times and take an average of the total actual cost (not the θ^{cost}-percentile). Also consider the risk of being "late," i.e., the total actual cost being greater than a given threshold. Plot the results and compare them.

[1] See http://tinyurl.com/sdagithub

7

Applications, Revisited

Now that we have reviewed a series of problem settings, we are going to pause and use these applications to illustrate in greater depth some of the modeling issues we touch on in Chapter 1.

Starting with the inventory problems in Section 1.3, we have now covered six classes of sequential decision problems. For each problem, we illustrated one or two strategies for making decisions:

Chapter 1) Inventory problems - We introduced sequential decision problems using variations of a simple inventory problem. Policies included order-up-to, and an adjusted forecast.

Chapter 2) Selling an asset - We had to decide when to sell a financial asset. Policies included variations of buy-low, sell-high.

Chapter 3) Adaptive market planning - This problem used a derivative-based stochastic search, where the sequential decision problem was choosing a stepsize, which we illustrated using simple parametric functions.

Chapter 4) Learning the best diabetes treatment - This is a classic active learning problem known as the multiarmed bandit problem. We designed policies based on parameterized optimization problems.

Chapter 5) Static stochastic shortest paths - We found an optimal solution of a particular version of a stochastic shortest path problem using a classical dynamic programming recursion that we could solve exactly, and then introduced a more complex stochastic version that we solved using approximate dynamic programming, exploiting a post-decision state variable.

Chapter 6) Dynamic stochastic shortest paths - Here we switch to a dynamic shortest path problem where estimates of the expected path costs evolve over time (in the static case in chapter 5, our estimates of the expected costs did not change). We used this to illustrate a basic deterministic lookahead policy, and a parameterized lookahead policy.

In Section 1.6, we introduced four classes of policies. In the applications we have reviewed so far, we have seen illustrations of each of the four classes. In this chapter, we are going to review the four classes in greater depth, and then we will return to our set of applications and identify the class for each of the suggested policies.

7.1 The four classes of policies

We first observe that the four classes of policies can be divided into two categories: the policy search class, and the lookahead class. Each of these can then be further divided into two classes, creating the four classes of policies. These are described in more detail below.

7.1.1 Policy search

The "policy search" class of policies involves searching over a set of functions for making decisions to find the function that works the best on average, using whatever objective is appropriate for the problem. Most of the time this will mean searching for the best value of a set of parameters that characterize a parameterized policy, but it also means that we may have to evaluate different parameterizations.

Policy-search policies can be divided between two classes:

- Policy function approximations (PFAs) - These are analytical functions that directly map a state to an action. Some examples are:
  - A parameterized function such as the buy-low, sell-high policy given in Equation (2.28), which we repeat here

$$X^{high-low}(S_t|\theta^{high-low}) = \begin{cases} 1 & \text{if } p_t < \theta^{low} \text{ or } p_t > \theta^{high}, \\ 1 & \text{if } t = T \text{ and } R_t^{asset} = 1, \\ 0 & \text{otherwise.} \end{cases}$$

    where $\theta^{high-low} = (\theta^{low}, \theta^{high})$. Other examples are the order-up-to policy we saw in Section 1.3.1, Equation (1.7) and the adjusted forecast policy in Section 1.3.2, Equation (1.15).
  - A linear function, such as

$$X^{\pi}(S_t|\theta) = \theta_0 + \theta_1\phi_1(S_t) + \theta_1\phi_1(S_t) + \ldots + \theta_F\phi_F(S_t)$$

    where $(\phi_f(S_t))$, $f = 1, \ldots, F$ is a set of features ("linear" means linear in the parameter vector θ - the features $\phi_f(S_t)$ can be highly nonlinear in S_t). For example, we might be trying to decide how much to bid to have a movie advertised on a website, and a feature might be the genre of the movie or the name of the lead actor or actress.

    Linear functions (also known as "affine policies") are popular, but note that you could not use a linear function to approximate step functions such as the buy-low, sell-high or order-up-to policies illustrated above.
  - Advanced functions such as locally linear functions or neural networks, although these typically have large numbers of parameters (the weights in a neural network) that need to be tuned.
- Cost function approximations (CFAs) - These are policies that require solving a parameterized optimization problem, where we may parameterize either the objective function or the constraints.

CFAs open the door to solving high-dimensional decision problems. Some examples are

– A simple example of a parameterized cost function approximation is the interval estimation policy we introduced in Equation (4.6) and repeat here

$$X^{IE}(S^n|\theta^{IE}) = \arg\max_{x\in\mathcal{X}} \left(\bar{\mu}^n_x + \theta^{IE}\bar{\sigma}^n_x\right).$$

– Parameterized optimization models - We saw this in Chapter 6 when we chose the θ-percentile of the link costs. This is a widely used heuristic in industry that has been overlooked by the research literature. Airlines use this idea to optimize the movement of their aircraft and crews in the presence of significant weather delays. Grid operators planning the scheduling of energy generators will insert reserve capacity to make sure that demand can be covered if a generator fails.

Both PFAs and CFAs have parameters that have to be tuned. The only difference is whether the policy involves an imbedded optimization problem or not. Both are exceptionally powerful and are widely used in different settings.

7.1.2 Lookahead approximations

Policies based on lookahead approximations are constructed by approximating the downstream costs (or rewards) from making a decision now, which are then considered along with the initial cost (or reward) of the initial decision.

- Policies based on value function approximations (VFAs) - These are policies that are based on Bellman's equation. The most basic form of Bellman's equation for deterministic problems problems was first presented in Chapter 5, Equation (5.2) as

  $$V_t(S_t) = \min_{x\in\mathcal{X}_s} \left(C(S_t,x) + V_{t+1}(S_{t+1})\right).$$

  There are many problems where the transition to S_{t+1} involves information (contained in W_{t+1}) that is not known at time t, which means S_{t+1} is a random variable at time t. In this case, we have to insert an expectation as we did in Equation (5.12) which gives us

$$V_t(S_t) = \min_{x \in \mathcal{X}_s} \left(C(S_t, x) + \mathbb{E}\{V_{t+1}(S_{t+1})|S_t, x\} \right).$$

In practice, we typically have to replace the value function $V_{t+1}(S_{t+1})$ with an approximation $\overline{V}_{t+1}(S_{t+1})$, as we did in Section 5.6.3. The field that studies these approximations goes under names such as approximate dynamic programming, reinforcement learning (which originated in computer science), and adaptive dynamic programming (the term used in the engineering controls community). In this case, the policy would be given by

$$X^\pi(S_t) = \arg\min_{x \in \mathcal{X}_s} \left(C(S_t, x) + \mathbb{E}\{\overline{V}_{t+1}(S_{t+1})|S_t, x\} \right).$$

If we use the post-decision state S_t^x, we can write our policy as

$$X^\pi(S_t) = \arg\min_{x \in \mathcal{X}_s} \left(C(S_t, x) + \overline{V}_t^x(S_t^x) \right),$$

which we illustrated in Section 5.6.2.

We used the deterministic shortest path problem to illustrate an application where value functions could be computed exactly. This can sometimes be done in stochastic problems, but in most applications, it has to be done approximately. The challenge is doing computations that are of sufficiently high quality to produce effective policies.

A popular approximation strategy for value functions is to use a linear model given by

$$\overline{V}_t^x(S_t^x|\theta^{VFA}) = \sum_{f \in \mathcal{F}} \theta_f^{VFA} \phi_f(S_t^x), \tag{7.1}$$

where $(\phi_f(S_t^x))_{f \in \mathcal{F}}$ is a user-defined set of features and θ^{VFA} is a set of parameters chosen using approximate dynamic programming algorithms.

We fit the linear model by collecting "observations" of the value $\hat{v}_t^n$ of being in state S_t^n in the nth. Let $\bar{\theta}^{VFA,n-1}$ be the estimate of θ^{VFA} after $n-1$ updates. There are methods that allow us to use $\hat{v}_t^n$ to easily update $\bar{\theta}^{VFA,n-1}$ and obtain $\bar{\theta}^{VFA,n}$. This gives us a VFA policy that we can write as

$$\begin{aligned} X_t^{VFA}(S_t|\theta^{VFA}) &= \arg\max_x \left(C(S_t,x) + \overline{V}_t^x(S_t^x|\theta^{VFA})\right). \\ &= \arg\max_x \left(C(S_t,x) + \sum_{f\in\mathcal{F}} \theta_f^{VFA}\phi_f(S_t^x)\right). \end{aligned} \quad (7.2)$$

Approximating value functions using linear models has been very popular, but there are virtually no theoretical guarantees on the quality of the resulting solution. Even worse, there is empirical evidence that the results can be quite poor. Yet, it remains popular because it is an easy way to "get a number."

Also popular today is to use neural networks (especially deep neural networks) to approximate a value function. Neural networks are attractive since they avoid the need to design the set of features $(\phi_f(S_t))$ for $f \in \mathcal{F}$. Caution has to be used, especially when we have to work with noisy observations of the value function, since the massive flexibility of neural networks can cause overfitting.

- Direct lookahead approximations (DLAs) - The first three classes of policies require finding some form of a functional approximation: the policy (for PFAs), the function being optimized (for CFAs), or the value of being in a downstream state (for VFAs). However, there are many problems where these functional approximations are just not possible.

  The "right" way to solve a DLA is to solve the true problem in the future, starting from the state S_{t+1} produced by starting in state S_t, taking action x_t, and then observing the random information W_{t+1}. The hard part is that in addition to modeling future uncertainties $W_{t+1}, W_{t+2}, \ldots,$, we also have to make optimal decisions $x_{t+1}, x_{t+2}, \ldots,$ each of which depend on the future state $S_{t+1}, S_{t+2}, \ldots,$ ..., which are random.

Although it is quite messy (and perhaps frightening), this policy means solving

$$X^*(S_t) = \arg\min_{x_t \in \mathcal{X}} \Bigg(C(S_t, x_t) + \\ \mathbb{E}_{W_{t+1}} \left\{ \min_{\pi} \mathbb{E}_{W_{t+2},\ldots,W_T} \left\{ \sum_{t'=t+1}^{T} C(S_{t'}, X^{\pi}(S_{t'})) | S_{t+1} \right\} | S_t, x_t \right\} \Bigg). \tag{7.3}$$

If we could compute Equation (7.3), we would have an optimal policy. It is quite rare that Equation (7.3) can be solved exactly. The basic stochastic shortest path problem in Chapter 5 is an example, but this is because uncertainty arises in a particularly simple way.

In most applications, we approach solving (7.3) by solving an approximate lookahead model. Instead of writing our sequence of states, decisions and information as

$$(S_0, x_0, W_1, \ldots, S_t, x_t, W_{t+1}, \ldots),$$

we create a simplified set of states, decisions and information for a model that we are solving at time t that we represent using

$$(\tilde{S}_{tt}, \tilde{x}_{tt}, \widetilde{W}_{t,t+1}, \ldots, \tilde{S}_{tt'}, \tilde{x}_{tt'}, \widetilde{W}_{t,t'+1}, \ldots),$$

where $\tilde{S}_{tt'}$ is typically a simplified state variable for the lookahead model we create when making a decision at time t, for time t' in the lookahead model. $\tilde{x}_{tt'}$ is our (possible simplified) decision created for time t' in the lookahead model, and $\widetilde{W}_{tt'}$ is the simplified information process at time t' in the lookahead model. Decisions $\tilde{x}_{tt'}$ are made using a *lookahead policy* $\tilde{X}_t^{\tilde{\pi}}(\tilde{S}_{tt'})$ which is typically a simplified policy chosen because it is easy to compute.

Our policy based on our approximate lookahead model would be written as

$$
X^{DLA}(S_t) = \arg\min_{x_t \in \mathcal{X}} \Bigg(C(S_t, x_t) +
$$

$$
\tilde{E}_{\tilde{W}_{t,t+1}} \left\{ \min_{\tilde{\pi}} \mathbb{E}_{\tilde{W}_{t,t+2},\ldots,\tilde{W}_{tT}} \left\{ \sum_{t'=t+1}^{T} C(\tilde{S}_{tt'}, \tilde{X}_t^{\tilde{\pi}}(\tilde{S}_{tt'})) | \tilde{S}_{t,t+1} \right\} | S_t, x_t \right\} \Bigg). \tag{7.4}
$$

We have already seen applications of this approach. For the dynamic shortest path problem in Chapter 6, Section 6.4.1, we turned to the widely used approach of solving a deterministic lookahead model, where we take the best estimate of what might happen in the future and solve a deterministic optimization problem. This approach ignores the effect of future uncertainties, but we introduced the idea in Section 6.4.2 of using a parameterized deterministic optimization problem. However, we have to tune the parameter.

The design of the lookahead policy $\tilde{\pi}$ (sometimes called the policy-within-a-policy) is highly problem dependent. In fact, we can use any of our four classes of policies. The key is that it has to be computationally simple, since we will have to compute it many times. Remember that the lookahead model does not have to be exact (in most cases we could never solve it if we tried to use an exact lookahead model). Instead, we are choosing approximations that we think will produce good decisions now, by approximating decisions that we *might* make in the future.

These four classes of policies (PFAs, CFAs, VFAs and DLAs) are universal, which is to say, any policy chosen for a sequential decision problem (*any* sequential decision problem) will be one of these four classes. However, these can also be building blocks for hybrid policies.

7.2 Models, revisited

In this section we are going to do a tour through the different applications, starting first with a review of the state variables. Then we are going to review the different policies, and classify the policies we have seen into the four classes.

7.2.1 State variables, revisited

There is considerable confusion in the academic literature about what is meant by a state variable, as evidenced by the noticeable absence of definitions of what a state variable is in books on dynamic programming, stochastic programming, and reinforcement learning.

The only exception to this pattern, which really stands out, is the optimal control literature where definitions of state variables are quite common. In the controls community, a state variable is commonly defined as "all the information we need at time t to model a system from time t onward." What is missing, however, is any description of precisely what information is needed to model the system from time t onward.

We define two versions of state variables (from Powell, 2022, Section 9.4):

Definition 7.2.1. A **state variable** is:

a) **Policy-dependent version** - A function of history that, combined with the exogenous information (and a policy), is necessary and sufficient to compute the cost/contribution function, the decision function (the policy), and any information required by the transition function to model the information needed for the cost/contribution and decision functions.

b) **Optimization version** - A function of history that is necessary and sufficient to compute the cost/contribution function, the constraints, and any information required by the transition function to model the information needed for the cost/contribution function and the constraints.

We need the two versions since if we have a system where we have specified the structure of a policy, we need to be sure we include any information needed by the policy. For example, we may have an inventory problem, where we consider two policies: one that uses a forecast of future demands, while the other just uses an order-up-to policy. While a forecast certainly seems relevant, if we are using an order-up-to policy, we are not using the forecast, and as a result it would not be in the state variable.

It helps to do a tour through our applications up to now and review the state variables for each one. For each application, we are going to summarize the state variable, which we might write as S_t or S^n depending on the setting, and we are going to classify the elements as physical state variables R_t, informational variables I_t, and belief state variables B_t.

Chapter 1 - This chapter introduced two inventory problems were also designed to bring out different flavors of state variables. The simple inventory problem in Section 1.3.1 was characterized by a state variable S_t that consists of just the inventory R_t^{inv} at time t. This problem is one of the most widely used applications for illustrating dynamic programming.

The more complex inventory problem in Section 1.3.2 required a state variable

$$S_t = (\underbrace{R_t^{inv}}_{R_t}, \underbrace{c_t}_{I_t}, \underbrace{f_{t,t+1}^D, \bar{\sigma}_t^D, \bar{\sigma}_t^f}_{B_t}).$$

This state variable illustrates all three classes of information in state variables: the physical state variables $R_t = R_t^{inv}$, other information $I_t = c_t$, and belief state variables $B_t = (f_{t,t+1}^D, \bar{\sigma}_t^D, \bar{\sigma}_t^f)$ where $(f_{t,t+1}^D, \bar{\sigma}_t^D, \bar{\sigma}_t^f)$ captures the forecasted mean and standard deviation of the error of the future demand $\hat{D}_{t+1}$, and the standard deviation in the change in forecasts from time t to $t+1$ (we assume that the change in forecasts have mean zero).

Chapter 2 - This chapter introduced a simple asset selling problem with state variable

$$S_t = (R_t^{asset}, p_t).$$

where the physical state variable R_t captures whether we are still holding the asset or not (it could have also held how many shares of stock we were holding), and the informational state $I_t = p_t$ is the price at which we sell the stock.

In Chapter 2, we introduced the idea of computing a smoothed estimate of the price of the asset using

$$\bar{p}_t = (1 - \alpha)\bar{p}_{t-1} + \alpha\hat{p}_t. \tag{7.5}$$

We then designed a policy that made decisions based on how much the price p_t deviated from this smoothed estimate. Now our state variable becomes

$$S_t = \big(\underbrace{R_t^{asset}}_{R_t}, \underbrace{(p_t, \bar{p}_t)}_{I_t}\big).$$

Now imagine that when we decide to sell our stock at time t, we sell at an unknown price p_{t+1} which evolves according to

$$p_{t+1} = \eta_0 p_t + \eta_1 p_{t-1} + \eta_2 p_{t-2} + \varepsilon_{t+1},$$

where ε_{t+1} is a mean 0 noise term. Now our state variable would look like

$$S_t = \big(\underbrace{R_t^{asset}}_{R_t}, \underbrace{(p_t, p_{t-1}, p_{t-2})}_{I_t}\big).$$

Chapter 3 - Here we described a gradient-based search algorithm that evolves according to a classical stochastic search iteration given by

$$x^{n+1} = x^n + \alpha_n \nabla_x F(x^n, W^{n+1}).$$

This procedure is a method for searching for the best value of x, but this is a sequential decision problem where the stepsize α_n is the decision. If we choose the stepsize with a deterministic formula such as $\alpha_n = 1/n$, then the "state" of our search procedure is

$$S^n = (x^n).$$

However, we might use an adaptive (stochastic) stepsize formula such as

$$\alpha_n = \frac{\theta}{\theta + N^n - 1} \tag{7.6}$$

where N^n is the number of times the gradient $\nabla_x F(x^n, W^{n+1})$ changes direction, then we need to know N^n, and our state variable becomes

$$S^n = (x^n, N^n).$$

Chapter 4 - Our diabetes problem is an instance of a pure learning problem, where we are trying to learn the true response μ_x of a patient to a drug. After trying several drugs, we might capture our belief using the state

$$S^n = \underbrace{(\bar{\mu}_x^n, \bar{\sigma}_x^n)_{x\in\mathcal{X}}}_{B^n},$$

where we assume that the true response $\mu_x \sim N(\bar{\mu}_x^n, (\bar{\sigma}_x^n)^2)$.

This belief model might work if we have a different belief for each patient, but presumably we start with a body of knowledge of how the drug works across all patients. We might capture this in an initial state

$$S^0 = (\bar{\mu}_x^0, \bar{\sigma}_x^0)_{x\in\mathcal{X}}.$$

Now imagine that the nth patient arrives with attributes a^n (gender, weight, smoking history, ...). The response of the patient to drug x would depend on both the drug as well as the attributes of the patient. This means that our state variable (that is, the information we have available to make the decision) consists of information we do not control (the attributes of the patient a^n), and information we do control (the choice of drug x^n). So we would write our state variable (the information we use to make the decision) as

$$S^n = (\underbrace{a^n}_{I^n}, \underbrace{(\bar{\mu}_x^n, \bar{\sigma}_x^n)}_{B^n})_{x\in\mathcal{X}},$$

where we decided to put a^n in our informational state variable I^n, and the variables $(\bar{\mu}_x^n, \bar{\sigma}_x^n)$ in the belief state variable B^n.

Chapter 5 - For our stochastic shortest path problem, we started with a basic problem where a traveler incurs a random cost when traversing a link, but only knows the mean and variance of the costs before making a decision at node i which link (i, j) to traverse. For this problem, the state of our traveler is simply the node N_t where he is located after traversing t links, giving us

$$S_t = N_t.$$

We then transitioned to a problem where the traveler at node i gets to see the actual costs $\hat{c}_{tij}$ that would be incurred if he were to travel over the link (i, j). With this additional information, the state variable becomes

$$S_t = \left(\underbrace{N_t}_{R_t}, \underbrace{(\hat{c}_{t,N_t,j})}_{I_t}{}_{j \in \mathcal{N}_i^+} \right).$$

Chapter 6 - We considered a dynamic shortest path problem where the estimated cost on link (i, j), $\bar{c}_{tij}$, evolves over time. That is, at time $t + 1$, we assume we are given an updated set of estimates that we would denote $\bar{c}_{t+1}$. Imagine that our traveler is at node $N_t = i$. The state of our system (for our traveler) would then be given by

$$S_t = (\underbrace{N_t}_{R_t}, \underbrace{\bar{c}_t}_{I_t}).$$

Now imagine that we show the traveler a path that we designate p_t which is the set of links that we are planning to get from his current node N_t to the destination. Let's say that we just updated the path, and have asked the traveler if he accepts the new path. If he says yes, the navigation system will continue to re-optimize, but will introduce a small bonus for staying with the latest path p_t that the traveler just accepted (this is done to prevent the system from chattering between two nearly equivalent paths).

If p_t is the most recently accepted path, then this is information that we need to make decisions in the future. In this case, our state variable becomes

$$S_t = (\underbrace{N_t}_{R_t}, \underbrace{(\bar{c}_t, p_t)}_{I_t}).$$

These decision problems have illustrated all three types of state variables: physical state variables R_t, informational state variables I_t, and belief state variables B_t. We have seen problems that have just R_t, or just B_t, and combinations with I_t such as (R_t, I_t) and (I_t, B_t), as well as all three (R_t, I_t, B_t). We emphasize that the distinction between R_t and I_t can sometimes be arbitrary, but there are so many problems that involve managing physical or financial resources (buying, selling, moving, modifying), with decisions that affect (or are constrained by) physical or financial resources, that we felt it was necessary to create a special class just for resources.

We think that there are many problems involving uncertainty that also involve learning, and may involve active learning since decisions may impact what we observe (as in the diabetes example). We suspect that as modelers become comfortable with including belief state variables in sequential decision problems that we will see them move often.

7.2.2 Policies, revisited

Our six application settings (and in some cases the extensions) were chosen to expose each of the four classes of policies. Below we review the different policies and identify the class to which they belong.

Chapter 1 - We introduced two inventory problems. One used an order-up-to policy of the form

$$X^\pi(S_t|\theta) = \begin{cases} \theta^{max} - R_t & \text{if } R_t < \theta^{min}, \\ 0 & \text{otherwise,} \end{cases} \tag{7.7}$$

while the second used a policy of bringing the inventory up to the forecasted demand plus a buffer

$$X^\pi(S_t|\theta) = \max\{0, f^D_{t,t+1} - R_t\} + \theta. \tag{7.8}$$

Both of these policies involve one (7.8) or two (7.7) tunable parameters. Both are analytical functions that do not have an imbedded optimization operator (min or max). These are the distinguishing characteristics of a policy function approximation (PFA).

Chapter 2 - This chapter addressed the problem of determining when to sell an asset. Several policies were suggested, but representative samples are the "sell-low" policy given by

$$X^{sell-low}(S_t|\theta^{low}) = \begin{cases} 1 & \text{if } p_t < \theta^{low} \text{ and } R_t = 1, \\ 1 & \text{if } t = T, \\ 0 & \text{otherwise.} \end{cases}$$

and the "tracking policy"

$$X^{track}(S_t|\theta^{track}) = \begin{cases} 1 & \text{if } p_t \geq \bar{p}_t + \theta^{track}, \\ 1 & \text{if } t = T, \\ 0 & \text{otherwise.} \end{cases} \tag{7.9}$$

Both of these policies are similar to our "order-up-to" inventory ordering policy in that they are parametric functions with tunable parameters, which means they are additional examples of policy function approximation (PFA). While this is hardly the only way to solve an asset selling problem, this class of policy is fairly popular on Wall St.

PFAs are popular in practice because of their simplicity and transparency, but it is important to keep in mind: *The price of simplicity is tunable parameters... and tuning is hard!.*

Chapter 3 - Adaptive market planning - This problem involves using a popular gradient-based search method (see Equation (7.6) where the stepsize α_n is the decision. If we had a deterministic problem, we would compute α_n by solving the one-dimensional optimization problem

$$\alpha_n = \arg\max_{\alpha \geq 0} \left(F(x^n + \alpha \nabla_x F(x))\right),$$

which is a form of direct lookahead approximation (DLA). However, when we have to deal with uncertainty, a one-dimensional search means we have to be able to compute the expectation $F(x) = \mathbb{E}F(x, W)$ which is generally not possible in practice. Instead, we might use a deterministic policy such as

$$\alpha_n^\pi(\theta) = \frac{\theta}{\theta + n - 1},$$

where we have written this as a parameterized function (that is, a form of PFA). We also illustrated an adaptive (state-dependent) policy given by Equation (7.6) where we replaced n with a counter N^n that counts how many times the gradient changes direction (or we could count how many times the objective function does not improve). We would write this policy as

$$\alpha_n^\pi(S^n|\theta) = \frac{\theta}{\theta + N^n - 1},$$

where our state S^n carries the information N^n.

Side note: PFA-style policies are universally used in stochastic gradient algorithms. While these may in fact be best, the reality is that no-one has even tried using the other three classes of policies. Might be worth a look.

Chapter 4 - Learning the best diabetes treatment - This is a pure learning problem which we have approached using the highly popular class of policies known as upper confidence bounding. Perhaps the best known UCB policy is given by

$$X^{UCB}(S^n|\theta^{UCB}) = \arg\max_{x\in\mathcal{X}} \left(\bar{\mu}_x^n + \theta^{UCB}\sqrt{\frac{\log n}{N_x^n}}\right). \quad (7.10)$$

Another variant which works very well was originally introduced as interval estimation which is given by

$$X^{IE}(S^n|\theta^{IE}) = \arg\max_{x\in\mathcal{X}} \left(\bar{\mu}_x^n + \theta^{IE}\bar{\sigma}_x^n\right). \quad (7.11)$$

Finally, a variant that was originally discovered in 1933 and then re-discovered a few years ago is Thompson sampling, which is given by

$$X^{TS}(S^n|\theta^{TS}) = \arg\max_{x\in\mathcal{X}} \hat{\mu}_x^n. \quad (7.12)$$

where $\hat{\mu}_x^n$ is randomly sampled from a normal distribution with mean $\bar{\mu}_x^n$ and variance $\theta^{TS}\bar{\sigma}_x^n$.

Note that all three policies ((7.10), (7.11) and (7.12)) share two features: an optimization operator (an $\arg\max_x$ for these policies) and a tunable parameter. These can be viewed as parameterized optimization problems, which belong to the class parametric cost function approximation (or CFA).

CFA policies are widely used in practice, but have received very little attention in the academic literature outside of the specific application of learning policies such as our diabetes application. We are going to see this idea applied in a very different setting in chapter 6.

Chapter 5 - Static stochastic shortest paths - Our first stochastic shortest path problem assumed that a traveler incurred stochastic costs, but these were learned only after traversing a link. This assumption allowed us to solve the problem as a deterministic shortest path problem, which is easily solved using Bellman's equation, giving us a policy that is given by

$$X_t^\pi(S_t = i) = \arg\min_{j \in \mathcal{N}_i^+} \left(\bar{c}_{tij} + V_{t+1}(S_{t+1} = j)\right). \tag{7.13}$$

where $S_t = N_t = i$ is the node where the traveler is located. The value functions $V_t(S_t)$ are computed by working backward in time, starting at $t = T$ where we set $V_T(S_T) = 0$ for all nodes S_T. This is a form of policy based on value function approximations, and this is a rare case where a VFA policy is actually optimal.

We then transitioned to a harder problem where a traveler is allowed to see the costs $\hat{c}_{tij}$ out of node $i = N_t$. For this problem, the state variable becomes $S_t = (N_t, (\hat{c}_{t,N_t,j},\ j \in \mathcal{N}_i^+))$. For this problem we had to approximate the value function using the post-decision state $S_t^x = N_t^x$ where N_t^x is the node we have chosen to go to after making our decision x_t when we are at node N_t. In this case our policy looked like

$$X_t^\pi(S_t = i) = \arg\min_{j \in \mathcal{N}_i^+} \left(\hat{c}_{tij} + \overline{V}_t^x(S_t^x)\right). \tag{7.14}$$

This again is a VFA-based policy, but this time it is no longer optimal since $\overline{V}_t^x(S_t^x)$ is an approximation we had to estimate from data. With some care, however, we can devise an asymptotically optimal policy.

Chapter 6 - Dynamic stochastic shortest paths - This is where we encounter a problem where the estimated costs on each link $\bar{c}_{tij}$ is evolving over time. So, at time t, $\bar{c}_t$ is the vector of estimated link costs, that becomes $\bar{c}_{t+1}$ the next time period. This means that our state variable transitions from $S_t = N_t$ which is just the node where the traveler is located, to $S_t = (N_t, \bar{c}_t)$ which is an extremely high-dimensional state variable. This is not a problem we can approach even with approximate dynamic programming (it is hard to envision a VFA built around this state variable).

Instead, we propose the idea of using a lookahead model, where we ignore the fact that as the traveler progresses over the network that the vector of estimated link costs $\bar{c}_t$ will evolve over time. Instead, we can assume it is fixed (and let's assume deterministic). This means we now have a lookahead model that is, in fact, a deterministic shortest path problem, but we have to remember that we are optimizing a deterministic lookahead model, which is a DLA policy. Of course, we know how to do this optimally, but an optimal solution to an approximate lookahead model is not an optimal policy!

Deterministic lookaheads are popular, but there is a way to make them even better, without making them more complicated. We introduced that idea when we suggested using the θ-percentile of the cost rather than the mean $\bar{c}_t$. Let $\tilde{c}_{tij}(\theta)$ be the θ-percentile of the cost on link (i, j) given what we know at time t. Now, solve a deterministic lookahead model using the costs $\tilde{c}_{tij}(\theta)$. Now we have a parameterized deterministic lookahead model, which is a hybrid of a parametric CFA and a DLA.

We have illustrated all four classes of policies, which leaves the question: How do you know which one to use? Sometimes it will seem obvious, such as finding the best path to a destination. For problems like this, a direct lookahead is a natural choice. But there are problems where any of the four classes is a viable candidate.

Two problems where we have successfully demonstrated all four classes are the inventory problems in Chapter 1, and the diabetes learning problem in Chapter 4. The key is to think carefully about all four classes of policies, rather than just focusing on one, which is what is happening so often today.

7.3 On the suboptimality of direct lookahead policies

An area of considerable confusion, even among experts, is the question of whether a direct lookahead policy is optimal or not, since all DLAs are themselves optimization problems of a dynamic model. We have to first remember what problem we are trying to solve.

Remember that we are always trying to solve a sequential decision problem that we write as a sequence of state variables (what we know when we make a decision), the decision variables (this is what we have to choose) and the exogenous information (information that arrived to the system after we make the decision). We can write this as

$$(S_0, x_0, W_1, S_1, x_1, W_2, \ldots, S_t, x_t, W_{t+1}, \ldots, S_T).$$

Every time we make a decision we receive a contribution $C(S_t, x_t)$ that may (and often does) depend on the dynamically changing information in the state variable S_t. Decisions are made with a method we call a policy $X^\pi(S_t)$, where we have now provided a number of examples. The state variable evolves according to the transition function

$$S_{t+1} = S^M(S_t, x_t = X^\pi(S_t), W_{t+1}),$$

where W_{t+1} is unknown when we make the decision x_t, but becomes known after we determine x_t so that we can calculate S_{t+1}. Our goal is to find the best policy $X^\pi(S_t)$, and we write this problem as

$$\max_\pi \mathbb{E}\left\{\sum_{t=0}^{T} C(S_t, X^\pi(S_t))|S_0\right\}. \tag{7.15}$$

Remember that we refer to the objective function (7.15), along with the transition function and our model of the exogenous information process $W_1, \ldots, W_T$ as the *base model.*

Above (in Equation (7.3)) we observed that the optimal policy is given by simply restating the objective in (7.15) starting at time t. For convenience we provide the policy again:

$$X^*(S_t) = \arg\max_{x_t \in \mathcal{X}} \Bigg(C(S_t, x_t) +$$
$$\mathbb{E}_{W_{t+1}} \left\{ \max_{\pi} \mathbb{E}_{W_{t+2},\ldots,W_T} \left\{ \sum_{t'=t+1}^{T} C(S_{t'}, X^{\pi}(S_{t'})) | S_{t+1} \right\} | S_t, x_t \right\} \Bigg). \tag{7.16}$$

The policy in Equation (7.16) is almost always computationally intractable. The process of "decision, information, decision, information, ..." invariably produces an explosive growth in the number of possible sample paths, even for toy problems.

The most common approach to solving sequential decision problems under uncertainty is to ignore uncertainty, and simply solve a sequence of deterministic lookahead models. This would give us a policy

$$X_t^{Det}(S_t) = \arg\max_{x_t} \left(C(S_t, x_t) + \max_{x_{t+1},\ldots,x_{t+H}} \sum_{t'=t+1}^{t+H} C(S_{t'}, x_{t'}) \right). \tag{7.17}$$

Typically, the feasible set of values for x_{t+1} depend on x_t (and so on into the future). It is natural to rewrite (7.17) as a single optimization problem rather than an optimization problem imbedded within an optimization problem, giving us

$$X_t^{Det}(S_t) = \arg\max_{x_t,\ldots,x_{t+H}} \sum_{t'=t}^{t+H} C(S_{t'}, x_{t'}). \tag{7.18}$$

We are all familiar with deterministic direct lookaheads such as (7.18); this is how our navigation systems are finding the best path to our destination, where the problem is typically reoptimized as we progress to reflect up-to-date traffic conditions (and whether we are actually following the recommended path).

It should be apparent that solving a deterministic approximation of the future to make decisions would not be an optimal policy. To obtain an optimal policy, we have to solve (7.16), which is simply not going to happen. Just the same, it is not unusual for people to fall in the trap of thinking that if they find an optimal solution to the deterministic lookahead model in (7.18), then the solution is, well, optimal. It an optimal solution to (7.18), but as a result of the deterministic approximation, the function $X^{Det}(S_t)$ in (7.18) *is not an optimal policy!*

There are substantial research communities who accept that a deterministic lookahead such as (7.18) is not optimal simply because it is deterministic, and the real problem is stochastic. However, the trap that people fall into is to replace the original sequential decision problem with a simplified stochastic model, which we can write as (starting at time t)

$$(\tilde{S}_{tt}, \tilde{x}_{tt}, \widetilde{W}_{t,t+1}, \ldots, \tilde{S}_{tt'}, \tilde{x}_{tt'}, \widetilde{W}_{t,t'+1}, \ldots),$$

which we introduced earlier. Here, we assume that we have created a simplified model in ways that make it solvable, but it is still a sequential decision problem. However, it is quite likely that we have simplified the state variables, decision variables, and/or the exogenous information, using a model that we create at time t.

Using our simplified lookahead model, we obtain a lookahead policy that we introduced earlier, but repeat again here as

$$\begin{aligned} X^{DLA}(S_t) = \arg\min_{x_t \in \mathcal{X}} \Bigg(& C(S_t, x_t) + \\ & \tilde{E}_{\widetilde{W}_{t,t+1}} \left\{ \min_{\tilde{\pi}} \mathbb{E}_{\widetilde{W}_{t,t+2}, \ldots, \widetilde{W}_{tT}} \left\{ \sum_{t'=t+1}^{T} C(\tilde{S}_{tt'}, \tilde{X}_t^{\tilde{\pi}}(\tilde{S}_{tt'})) | \tilde{S}_{t,t+1} \right\} | S_t, x_t \right\} \Bigg). \end{aligned} \tag{7.19}$$

This is still a stochastic lookahead model, which means it is probably hard to solve, but the hope is that we have introduced simplifications that make it at least tractable. Note that there are entire books written on different strategies for approximating stochastic lookahead policies (as given in (7.16)) with approximate lookahead models, giving us the policy in (7.19). However, optimizing a stochastic lookahead model using an approximate model (and the deterministic lookahead is simply one example of an approximation), even if it requires sophisticated algorithms (and it often does), *is still not an optimal policy!*

For readers looking for an in-depth discussion of direct lookahead policies, please see Powell (2022, Chapter 19).

7.4 Online vs. offline objectives

There are two perspectives for evaluating the performance of a policy:

Online learning - There are many settings where we have to learn as we go in the field. For example, we might be trying to learn the best price for a product, the best path through a congested city, or the best drug for a patient to lower blood pressure. In each of these cases, we want to do as well as we can given what we know, but we are still learning so we can make better decisions in the future. This means we need to maximize the *cumulative reward* over time (or iterations) so we capture how well we are doing while we are learning.

Offline learning - In other cases, we can learn in a laboratory setting, which might be a physical lab (to test different materials or observe the performance of a drug on mice), a computer simulator, or even a field setting which is being run as a test market (to evaluate a product) or a clinical trial (to test drugs). If we are learning in a laboratory environment ("offline") then we are willing to perform trial and error without worrying about how well we do. Instead all we care about is the quality of the solution in the end, which means we want to optimize the *final reward.*

A word of caution about the terms online and offline. In the machine learning community, "offline" refers to the estimation of models using a single, batch dataset. By contrast, online learning is used to refer to fully sequential settings where data arrives over time. This is typically in field situations where data is created by some exogenous process (such as observing patients arriving to a doctor's office), which is the same setting we assume when using the term "online." However, in machine learning "online" would still be used to refer to an iterative algorithm being used in a simulation.

There are entire fields dealing with sequential decision problems that are separated by whether they focus on final reward or cumulative reward. For example, communities that work on "stochastic search" tend to focus on final reward, while the communities that work on "multiarmed bandit problems" (a form of stochastic search problem) generally optimize cumulative reward. The reality is that you can use the same policy for either objective, but you have to tune it for the objective you choose.

7.4.1 Online (cumulative reward) optimization

It is typically the case when doing policy search that we have a parameterized policy that we can write as $X^\pi(S_t|\theta)$. The decision $x_t = X^\pi(S_t|\theta)$ might be the price of a product, the choice of a blood-pressure drug or the bid placed to maximize ad-clicks. In all these cases, we have to learn as we go, which means we need to maximize performance as we are learning.

Let $C(S_t, x_t)$ be our performance metric (revenue, reduction in blood-pressure, or net revenue from ad-clicks). We want to find θ that produces the policy $X^\pi(S_t|\theta)$ that solves the optimization problem

$$\max_\theta F^\pi(\theta) = \mathbb{E}_{S_0}\mathbb{E}_{W_1,\ldots,W_T|S_0}\left\{\sum_{t=0}^{T} C(S_t, X^\pi(S_t|\theta))|S_0\right\}, \quad (7.20)$$

where $S_{t+1} = S^M(S_t, X^\pi(S_t|\theta), W_{t+1})$. The expectation in (7.20) is over all possible realizations of $W_1, \ldots, W_T$, as well as over the possible values of uncertain parameters (such as uncertain initial beliefs about market responses or how someone responds to a drug) that are contained in the initial state S_0.

Equation (7.20) is an example of an "online" or "cumulative reward" objective function, since we want to maximize the sum of all the rewards over some horizon. This is of particular interest in online learning problems where we have to learn the performance, such as revenue from a price or the performance of a drug for a particular patient, which means balancing the process of learning while also trying to do as well as we can.

7.4.2 Offline (final reward) optimization

In offline settings, we typically have a budget of N experiments. A classical (if inappropriate) problem that is often used to illustrate offline, derivative-free learning is the newsvendor problem which we addressed in Chapter 3. As a reminder, the newsvendor problem is written as

$$F(x) = \mathbb{E}_W(p\min\{x, W\} - cx), \tag{7.21}$$

where x is the quantity of resource we order at a unit cost c, which is then used to meet the demand W (which is unknown when we choose x). We assume the distribution of W is unknown.

Let $x^n = X^\pi(S^n|\theta)$ be our choice of x given what we know which is captured by S^n, where our policy $X^\pi(S^n|\theta)$ depends on one or more parameters in θ (as we illustrated in the policies introduced in Section 1.3). After we implement x^n, we observe W^{n+1}, update S^{n+1} and then repeat the process. After N iterations, we obtain a final design we denote $x^{\pi,N}(\theta)$.

We now have to evaluate our final design $x^{\pi,N}(\theta)$. To perform this evaluation, we have to consider two, and possibly three, sources of uncertainty. The first is that we may have uncertainty in unknown parameters such as the mean of W. For example, W might come from a Poisson distribution with mean μ, and we may assume that $\mu \in \{\mu_1, \ldots, \mu_K\}$ where $p_k = Prob[\mu = \mu_k]$. The distribution $(p_k)_{k=1}^K$ is contained in the initial state S_0.

Then we have the random arrivals of demands $W^1, \ldots, W^N$ which would be sampled from a Poisson distribution with mean μ. We use these observations, and the policy $X^\pi(S^n|\theta)$, to compute $x^{\pi,N}(\theta)$. It is important to recognize that $x^{\pi,N}(\theta)$ is a random variable that depends on any information in S^0 (regardless of whether it is deterministic or random).

Once we have computed $x^{\pi,N}(\theta)$, we have to run a final set of simulations to evaluate how well it works. We introduce a new random variable, $\widehat{W}$, to represent samples of W used for evaluating our final design.

This notation allows us to write our objective function for offline learning as

$$\max_\theta F^\pi(\theta) = \mathbb{E}_{S^0}\mathbb{E}_{W^1,\ldots,W^N|S^0}\mathbb{E}_{\widehat{W}|S^0}F(x^{\pi,N}(\theta), \widehat{W}). \tag{7.22}$$

We emphasize that we write expectations simply as a way of indicating that we have to average over random information. We address the problem of computing these expectations next.

7.4.3 Policy evaluation

The objective functions for cumulative reward (given in (7.20)) and final reward (given in (7.22)) were both written using expectations, which is our way of saying that we are averaging over whatever is random. This is nice to write mathematically, but these are virtually never computable.

Whenever we have to take an expectation, it helps to assume that we are going to estimate the expectation through sampling. We first illustrate how to do this for the cumulative reward objective as given in (7.20). Here, we might have an uncertain quantity in the initial state S_0, such as uncertainty in how a market responds to price, the production of methane by an oil well, or how a patient might respond to a drug. Then, we have the exogenous information $W_1, \ldots, W_T$, which could be observations of sales, the change in atmospheric temperatures, or how a patient responds to medication.

Let ω be a sample realization of all of these uncertain quantities. Assume that we generate a set of samples of all of these uncertain quantities and store them in a set $\Omega = \{\omega^1, \ldots, \omega^K\}$. So, anytime we write $W_t(\omega)$, this is a sample realization of what we observe at time t. If we are using a policy $X^\pi(S_t|\theta)$, then we would follow the sample path of states $S_t(\omega)$, actions $x_t(\omega) = X^\pi(S_t(\omega)|\theta)$ and exogenous information $W_{t+1}(\omega)$ governed by our transition function

$$S_{t+1}(\omega) = S^M(S_t(\omega), x_t(\omega), W_{t+1}(\omega)).$$

Using our set of sample observations Ω, we can approximate our expectation $F^\pi(\theta)$ using

$$\overline{F}^\pi(\theta) \quad = \quad \frac{1}{K}\sum_{k=1}^{K}\sum_{t=0}^{T} C(S_t(\omega^k), X^\pi(S_t(\omega^k)|\theta)). \qquad (7.23)$$

If we are using a final reward objective, we need to first estimate $x^{\pi,N}(\theta)$. If we follow sample path ω, then we would write our final design as $x^{\pi,N}(\omega|\theta)$, where ω captures everything we used to perform the training given by $(S_0(\omega), W_1(\omega), \ldots, W_T(\omega))$.

We then need to evaluate our design $x^{\pi,N}(\omega|\theta)$ using the testing data captured in $\widehat{W}$. Let ψ be a sample realization of $\widehat{W}$, and just as we assumed that we have a sample set Ω for ω, let's assume we create a set of sample outcomes of $\widehat{W}$ given by $\Psi = \{\psi^1, \ldots, \psi^L\}$. Keep in mind that $\widehat{W}$ represents any simulated information that we need to evaluate our design $x^{\pi,N}$. It may be a set of random variables (patient attributes, weather, market conditions), and it may even represent information that evolves over time. In other words... anything.

Now we would write the estimate of the performance of the policy in a final reward setting as

$$\overline{F}^\pi(\theta) = \frac{1}{K}\frac{1}{L}\sum_{k=1}^{K}\sum_{\ell=1}^{L} F(x^{\pi,N}(\omega^k), \widehat{W}(\psi^\ell)). \tag{7.24}$$

7.4.4 Bringing them together

Equation (7.20) illustrates an online, or cumulative reward, objective function where we have to maximize total performance during the learning process. Equation (7.22) illustrates the offline, or final reward, objective function where we have to search for the best design that will work best on average after we fix the design. What is important at the moment is that both problems involve solving

$$\max_\theta F^\pi(\theta), \tag{7.25}$$

where $F^\pi(\theta)$ is an unknown function which we can sample in a noisy way.

We can expand the objective in (7.25) to include a search over different classes of policies. Let $\mathcal{F}$ represent the set of all possible types of policies, including the major classes (PFAs, CFAs, VFAs and DLAs), as well as different functions within each of these classes. Then let Θ^f be the set of all possible parameter vectors θ that correspond to whatever policy class $f \in \mathcal{F}$ that we have chosen. In this case, we can write our optimization problem as

$$\max_{\pi=(f\in\mathcal{F},\theta\in\Theta^f)} F^\pi(\theta). \tag{7.26}$$

In practice, we tend to choose the class of policy $f \in \mathcal{F}$ using intuition and an understanding of the structure of the problem, but this is not always obvious. We urge readers to be ready to use intuition and common sense, but be aware of all four classes. This does not mean that you have to test all four classes, but you should be ready to defend why you made the choice you did.

We next turn to the problem of optimizing over θ, which we assume is continuous and, in most cases, vector-valued. There are two broad classes of search methods that we can apply to finding θ: derivative-based, and derivative-free.

7.4.5 Dependence on the initial state

Regardless of whether we are using a cumulative reward objective (such as Equation (7.20)) or final reward objective (such as Equation (7.22)), our optimization of θ will depend on the initial state S_0 (or S^0). This means that the solution we obtain from trying to solve this optimization problem (ideally the optimal solution) depends on S_0.

The initial state S_0 (or S^0 if appropriate) contains any and all information that affects the behavior of the system in any way. It may include deterministic parameters, distributions about uncertain parameters, and even the starting position of the search algorithm.

The dependence of optimal solutions on the information in S_0 is widely overlooked in the algorithmic literature. It would be nice if we could compute the function $\theta(S_0)$ to capture this dependence, but estimating this function is intractable. This means if S_0 changes, we have to reoptimize θ. That would be fine, except that there are many situations where S_0 changes, and we do not reoptimize θ simply because it can be quite difficult.

This is something that the reader should be aware of.

7.5 Derivative-based policy search

Assume that we are trying to solve the problem

$$\max_{\theta} F(\theta), \tag{7.27}$$

where $F(\theta)$ is some parametric function in θ. Further assume that θ is a vector and that we can compute the gradient

$$\nabla_\theta F(\theta) = \begin{pmatrix} \frac{\partial F(\theta)}{\partial \theta_1} \\ \frac{\partial F(\theta)}{\partial \theta_2} \\ \vdots \\ \frac{\partial F(\theta)}{\partial \theta_K} \end{pmatrix}.$$

In practice, computing derivatives exactly is often not possible.

A useful method for handling higher-dimensional parameter vectors is *simultaneous perturbation stochastic approximation* (or SPSA) developed by Spall (2003) which approximates gradients as follows. Let $Z_p, p = 1, \ldots, P$, be a vector of zero-mean random variables, and let Z^n be a sample of this vector at iteration n. We approximate the gradient by perturbing x^n by the vector Z using $x^n + \eta^n Z^n$ and $x^n - \eta^n Z^n$, where η^n is a scaling parameter that may be a constant over iterations, or may vary (typically it will decline). Now let $W^{n+1,+}$ and $W^{n+1,-}$ represent two different samples of the random variables driving the simulation (these can be generated in advance or on the fly). We then run our simulation twice: once to find $F(x^n + \eta^n Z^n, W^{n+1,+})$, and once to find $F(x^n - \eta^n Z^n, W^{n+1,-})$. The estimate of the gradient is then given by

$$\nabla_\theta F(\theta^n, W^{n+1}) \approx \begin{bmatrix} \frac{F(x^n+\eta^n Z^n, W^{n+1,+}) - F(x^n-\eta^n Z^n, W^{n+1,-})}{2\eta^n Z_1^n} \\ \frac{F(x^n+\eta^n Z^n, W^{n+1,+}) - F(x^n-\eta^n Z^n, W^{n+1,-})}{2\eta^n Z_2^n} \\ \vdots \\ \frac{F(x^n+\eta^n Z^n, W^{n+1,+}) - F(x^n-\eta^n Z^n, W^{n+1,-})}{2\eta^n Z_P^n} \end{bmatrix}. \quad (7.28)$$

Note that the numerator of each element of the gradient in Equation (7.28) is the same, which means we only need two function evaluations: $F(x^n + \eta^n Z^n, W^{n+1,+})$ and $F(x^n - \eta^n Z^n, W^{n+1,-})$. The only difference is the Z_p^n in the denominator for each dimension p (that is the magic of SPSA).

A brief word of caution about the "magic" of SPSA is that the gradients can be quite noisy. For this reason, a common strategy is to run multiple simulations (called mini-batches in the literature) of each perturbed simulation and average them. The appropriate size of the mini-batches depends on the problem characteristics, so anticipate spending some time tuning this parameter.

However we compute the gradient, our search algorithm (which we saw in Chapter 3), is given by

$$\theta^{n+1} = \theta^n + \alpha_n \nabla_\theta F(\theta^n, W^{n+1}). \tag{7.29}$$

We now have to choose a policy for the stepsize α_n, which we have discussed in Chapter 3, but see Powell (2022, Chapter 6) for a thorough discussion of stepsize policies. We remind the reader that as we learned in Chapter 3 that a stochastic gradient algorithm is itself a sequential decision problem (as we saw in Chapter 3).

7.6 Derivative-free policy search

Derivative-free policy search is simply another example of a sequential decision problem that is the focus of this entire volume, with the main difference that the only state variable will be the belief about the function we are maximizing (which is the same as our diabetes application in Chapter 4).

We can form beliefs using any of the following:

Lookup tables - Assume that we can discretize the set of possible values of θ into a set $\Theta = \{\theta_1, \ldots, \theta_K\}$. Define a random variable $\mu_\theta = F(\theta) = \mathbb{E}F(\theta, W)$ which is a random variable because we do not know $F(\theta)$ (or μ_θ). Assume that we can run experiments to sample $\hat{F}^{n+1} = F(\theta^n, W^{n+1})$. We can use these samples to create estimates $\bar{\mu}^n_\theta$ for each discrete value of $\theta \in \Theta$. This would be a lookup table belief model.

The simplest belief model is a lookup table with independent beliefs, which we first saw in Chapter 4, Section 4.2.1 with our diabetes application. Let $\bar{\mu}^n_\theta$ be our estimate of $\mathbb{E}F(\theta)$ for some $\theta \in \{\theta_1, \ldots, \theta_K\}$ after n samples (across all experiments). Let $\bar{\sigma}^n_{\theta_k}$ be the standard deviation of the estimate $\bar{\mu}^n_{\theta_k}$ and let $\beta^n_{\theta_k}$ be the precision given by

$$\beta^n_{\theta_k} = \frac{1}{(\bar{\sigma}^n_{\theta_k})^2}.$$

Parametric model - We might estimate a linear model of the form

$$F(\theta|\eta) \approx \eta_0 + \eta_1\phi_1(\theta) + \eta_2\phi_2(\theta) + \ldots$$

where $\phi_f(\theta)$ are features computed from the vector θ, which might consist of terms such as θ, θ^2, or $\ln\theta$. Note that a "linear model" means that it is linear in the coefficients η; the features $\phi_f(\theta)$ may be nonlinear functions of θ.

While linear models are popular, they are likely going to be little more than a local approximation. This is not a problem if we fit the linear model around the right point. The problem is finding the right point!

Nonparametric models - There are many ways to create nonparametric models. Perhaps the most common would be to create an estimate $\bar{\mu}_\theta$ by averaging over $\bar{\mu}_{\theta_1}, \ldots, \bar{\mu}_{\theta_K}$ for values θ_k that are close to θ. We are not going to draw on nonparametric methods in this book, primarily because they are data intensive and somewhat clumsy to use.

Powell (2022, Chapter 3) describes a number of methods for recursively estimating functions, covering several belief models for lookup tables, linear models, and nonlinear models. The chapter also covers both Bayesian and frequentist models. We already saw the recursive equations for a lookup table for our diabetes example, where the updating was given by Equations (4.1)-(4.2) (these equations assume a Bayesian belief model). Later we will illustrate the recursive updating of linear and nonlinear models.

- State variables - This would be the belief about $\mathbb{E}F(x, W)$, which we can write as $S^n = B^n$. How we store the belief state depends on whether we are using lookup tables (and what type of lookup table), linear or nonlinear models. If we are using our lookup table belief model, we would use

$$B^n = (\bar{\mu}_\theta^n, \beta_\theta^n), \ \theta \in \{\theta_1, \ldots, \theta_K\}.$$

- Decision variable - The decision is the choice θ^n of what value of θ to evaluate the function to obtain $\hat{F}^{n+1} = F(x^n, W^{n+1})$. We make our choice $\theta^n = \Theta^\pi(S^n)$ using a policy $\Theta^\pi(S^n)$ that we need to design.

- Exogenous information - We normally think of the exogenous information as being the sampled observation $\hat{F}^{n+1} = F(\theta^n, W^{n+1})$. Let β_θ^W be the precision (one over the variance) of the noise from observing the function $F(\theta^n, W^{n+1})$ at a point θ.

- Transition function - The transition function comes in the form of the recursive equations for updating our beliefs. For example, the updating Equations (4.1)-(4.2) for the lookup table beliefs in Chapter 4 where we were searching for the best diabetes medication. Again using our lookup table belief model, the transition function would be

$$\bar{\mu}_\theta^{n+1} = \frac{\beta_\theta^n \bar{\mu}_\theta^n + \beta_\theta^W W_\theta^{n+1}}{\beta_\theta^n + \beta_\theta^W}, \tag{7.30}$$

$$\beta_\theta^{n+1} = \beta_\theta^n + \beta^W. \tag{7.31}$$

- Objective function - It is most common that we would use a simulator offline to search for the best value of θ, which means a final-reward objective. Let $\theta^{\pi,N}$ be our the best value of the parameter vector θ derived from our belief model after N samples. For example, if we were using a lookup table belief model, and we obtain estimates $\bar{\mu}_\theta^N$ for each choice θ after N experiments, we would choose

$$\theta^{\pi,N} = \arg\max_{\theta\in\Theta} \bar{\mu}_\theta^N,$$

where the superscript "π" in $\theta^{\pi,N}$ reflects the type of search policy π used when estimating $\bar{\mu}_x^N$. The optimization problem to search for the best policy π would be written

$$\max_\pi \mathbb{E}_{S^0}\mathbb{E}_{W^1,\ldots,W^N|S^0}\mathbb{E}_{\widehat{W}|S^0} F(\theta^{\pi,N}, \widehat{W}).$$

Remember that we calculate expectations using simulation, as we showed in Section 7.4.3, Equation (7.23) for cumulative reward, or (7.24) for final reward.

This leaves us with the question: How do we design the search policy $\Theta^\pi(S^n)$? We hope that it does not come as a surprise that we can choose from any of the four classes of policies. All four classes are discussed in depth in Powell (2022, Chapter 7), but we would also refer to the discussion of policy search in Powell (2022, Chapter 12).

For our purposes, we are going to illustrate two policies that are relatively simple and natural.

- Interval estimation policies for lookup table belief models - We reviewed several policies for the diabetes setting in Section 4.4, but one that is particularly effective is the interval estimation policy, given by

$$\Theta^{IE}(S^n|\theta^{IE}) = \arg\max_{\theta\in\Theta}\left(\bar{\mu}^n_\theta + \theta^{IE}\bar{\sigma}^n_\theta\right). \tag{7.32}$$

  Now we have to search for the best value of θ^{IE}. Think of this as a sequential decision problem (finding the best θ^{IE}) to solve a sequential decision problem (to find the best θ for our policy $X^\pi(S_t|\theta)$). We note that in the search community, the tuning of the parameter θ^{IE} is typically overlooked in the research literature, but practitioners are aware that it has to be done.

- Classic response surface methods - For problems where x is continuous, it makes sense to fit a parametric model to the function $\mathbb{E}F(x, W)$. While it may be tempting today to use neural networks, keep in mind that these are high dimensional models that tend to overfit any noise. There are many problems where $F(x, W)$ is expensive; for example, it might be a computer simulation that could take an hour or more, or it could require field experiments.

  For these reasons, a popular strategy is to use a linear model of the form

$$\overline{F}(x) = \sum_f \theta_f \phi_f(x),$$

  where $\phi_f(x)$ is a feature extracted from the input vector x. Assume we have run n experiments using inputs $x^0, \ldots, x^{n-1}$ from which we have observed the responses $\hat{F}^1, \ldots, \hat{F}^n$. From this data, we can use the techniques of linear regression to fit our linear model with estimates $\theta \approx \bar{\theta}^n$, giving us the approximation

$$\overline{F}^n(x) = \sum_f \bar{\theta}^n_f \phi_f(x).$$

The response surface literature has long used the greedy strategy of optimizing $\overline{F}^n(x)$ to find the next point to observe, which means we would write

$$x^n = \arg\max_x \overline{F}^n(x). \tag{7.33}$$

While this idea is intuitively appealing, it turns out that using what appears to be the best estimate of the optimum based on our approximation $\overline{F}^n(x)$ is actually a terrible way to learn the best approximation of $\mathbb{E}F(x)$.

Figure 7.1 illustrates the challenges of learning a parametric function. Here we show three linear demand response curves giving demand as a function of price with the form $D(p) = \theta_0 - \theta_1 p$. Our goal is to maximize revenue $R(p) = pD(p)$. Figure 7.1(a) shows three possible demand curves and the corresponding revenue curves.

It is tempting to want to quote prices that optimize the revenue as we do in Figure 7.1(b), but this produces a set of points in a ball that makes it hard to estimate the demand curve. The best way to estimate the demand curve is to quote prices near the extremes as we do in Figure 7.1(c), but the revenue is very low at these points, so we do not make money while we are learning.

A good approach is to test points on the "shoulders" which means not at the optimum, but not too far away, as we do in Figure 7.1(d), a behavior we accomplish with a policy we describe next.

- Response surface methods with perturbation - A policy that overcomes the exploration vs. exploitation tradeoff illustrated in Figure 7.1 uses a one-step lookahead policy called the knowledge gradient that chooses x^n to be the value that produces the maximum value of information. It turns out the points that maximize the value

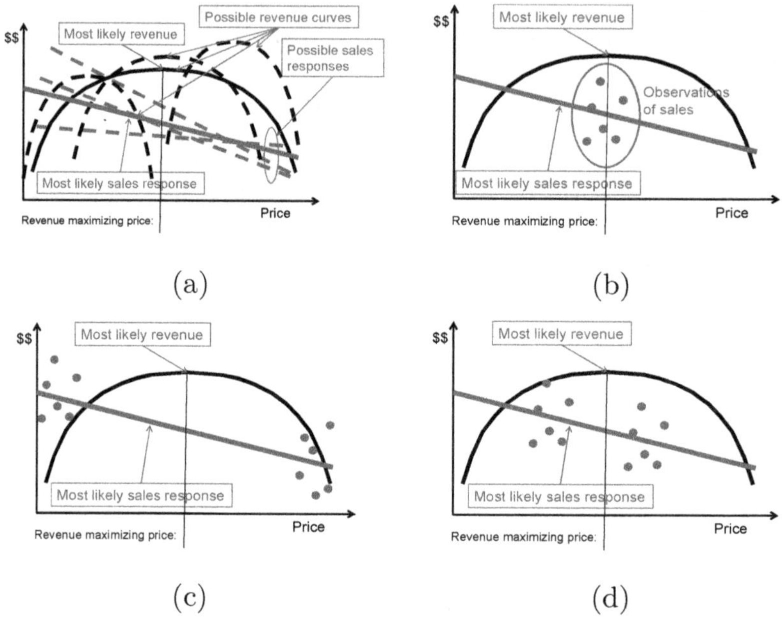

Figure 7.1: Actively learning a demand response function: (a) Three possible sales response lines and corresponding revenue curves, (b) Observed price-sales combinations if we use prices that appear to maximize revenue, (c) Observing extreme prices (high and low) to improve learning of sales response, and (d) Balancing learning (observing away from the middle) and earning (observing prices near the middle). From Powell (2022).

of information give us the sampling pattern illustrated in Figure 7.1(d).

The knowledge gradient is too complex for our presentation here, but there is a very simple way of getting the same behavior. Instead of taking our current estimate of the apparent optimum x^n as we do in (7.33), we perturb it by an amount ρ, but there are two ways of doing this:

An optimum-deviation policy - The idea here is to pick a point x^n that is a distance ρ from the optimum $\bar{x}^n = \arg\max_x \overline{F}^n(x|\bar{\theta}^n)$. If x is a $k-$dimensional vector, this deviation can be created by sampling k normally distributed random variables $Z_1, \ldots, Z_K$, each with mean 0 and variance 1, and then normalizing them so that

$$\sqrt{\sum_{k=1}^{K} Z_k^2} = \rho.$$

Let $\bar{Z}^n$ be the resulting $k-$dimensional vector. Now compute the sampling point using

$$x_k^n = \bar{x}_k^n + \bar{Z}_k^n.$$

Note that in one dimension, we would have $\bar{Z}^n = \pm\rho$.

An excitation policy - Here we again generate a $k-$dimensional perturbation vector Z^n, where each element has mean 0 and variance 1, and then set

$$x_k^n = \bar{X}_k^n + \rho Z_k^n.$$

While the optimum-deviation policy forces x^n to be a distance ρ from the optimum $\bar{x}^n$, an excitation policy simply introduces a random perturbation with mean 0, which means the most likely point to sample is the optimum of $\bar{f}^n(x|\bar{\theta}^n)$.

We suggest that the optimum-deviation policy is better suited for offline, final-reward objectives, while the excitation policy is better when we are in an online setting optimizing cumulative reward.

7.7 What did we learn?

- We review the four classes of policies.
- We use all the applications introduced in the previous six chapters to contrast the different styles of state variables, and to illustrate each of the four classes of policies.
- We also describe final reward and cumulative reward objectives. Final reward objectives arise when we are making observations in an experimental setting (in the lab or the field), where we only care about the performance of the final design. Cumulative reward objectives are used when we are learning while doing.

- We describe both derivative-based and derivative-free search methods for doing parameter search.
- We note that both derivative-based and derivative-free stochastic search are both sequential decision problems.

7.8 Exercises

Review questions

7.1 — What distinguishes PFAs from the other three classes of policies?

7.2 — What distinguishes VFAs and DLAs from PFAs and CFAs?

7.3 — In Chapter 1, the state variable in Section 1.3.2 for the more complicated inventory problem consists of physical state variables R_t, informational state variables I_t, and belief state variables B_t. What distinguishes a belief state variable from an informational state variable?

7.4 — What is the difference in the objective functions for online and offline learning?

7.5 — We described derivative-based and derivative-free stochastic search as sequential decision problems. Which one of these two strategies uses a belief state, and why is this necessary?

Problem solving questions

7.6 — Figure 7.2 shows a deterministic graph, where we are trying to find a path from node 1 to node 11 using different objectives.

a) If our traveler simply wants to minimize the total travel time from node 1 to 11, and has currently traversed the path 1-2-6-9, what is her state?

b) Now assume that our traveler has to arrive at node 11 by time 45. If she arrives after time 45, she is assessed a penalty equal to the square of the delay. What is the state of the traveler who has followed path 1-2-6-9 up to now?

c) What is the state if the traveler who has followed path 1-2-6-9 and wants to minimize the second highest cost on any of the links along her path?

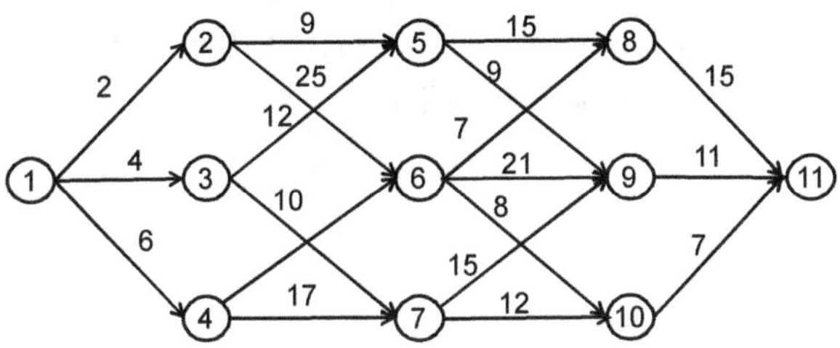

Figure 7.2: A deterministic graph.

7.7 — Real story: a finance technology company ("fintech") has an algorithmic trading system for high-frequency trading. At time t, when an asset is trading at price p_t, they estimate whether the price is going up or down using a series of forecasts of how the price might change over a rolling horizon within the day. Here, time is measured in 15-minute increments. Let $f_{tt'}$ be the estimated price of the asset at time t' made given the information at time t. Now create an estimated price using

$$\bar{f}_t(\theta) = \sum_{t'=t+1}^{t+H} \theta_{t'-t} f_{tt'},$$

where $\theta = (\theta_1, \theta_2, \ldots, \theta_H)$ is the vector of weights for each 15-minute increment for six hours into the future (24 increments). Let $x_t = 1$ indicate a decision to sell at time t, $x_t = -1$ is a decision to buy, and $x_t = 0$ is to hold, where the policy is

$$X^\pi(S_t|\theta) = \begin{cases} +1 & \text{if } \bar{f}_t(\theta) \geq p_t + 1.0, \\ 0 & \text{if } p_t - 1.0 < \bar{f}_t(\theta) < p_t + 1.0, \\ \text{-1} & \text{if } \bar{f}_t(\theta) \leq p_t - 1.0. \end{cases} \tag{7.34}$$

The challenge is to optimize the weight vector θ.

a) At time t, what is the state of this system?

b) What class of policy would $X^\pi(S_t|\theta)$ fall in? Explain.

c) Assume that you can simulate the policy using historical data in a simulator. Let $F(\theta)$ be the expected performance of the policy given parameter vector θ. Write out this objective assuming that you are going to simulate the policy using a single sample of history.

d) Describe how to compute a numerical derivative using your simulator. Just write the numerical derivative for a single element θ_τ.

8

Energy Storage I

8.1 Narrative

New Jersey is looking to develop 3,500 megawatts (MW) of off-shore wind generating power. A challenge is that wind (and especially offshore wind) can be highly variable. The effect of this variability on the power grid is magnified by the property that wind power (over intermediate ranges) increases with the cube of wind speed. This variability is depicted in Figure 8.1.

Energy from wind power has become popular in regions where wind is high, such as the midwestern United States, coastal regions off of Europe, the northeast of Brazil, and northern regions of China (to name just a few). Sometimes communities (and companies) have invested in renewables (wind or solar) to help reduce their carbon footprint and minimize their dependence on the grid.

It is quite rare, however, that these projects allow a community to eliminate the grid from their portfolio. Common practice is to let the renewable source (wind or solar) sell directly to the grid, while a company may purchase from the grid. This can be useful as a hedge since the company will make a lot of money during price spikes (prices may jump from \$20 per megawatt-hour (mwh) to \$300 per mwh or more) that offsets the cost of purchasing power during those periods.

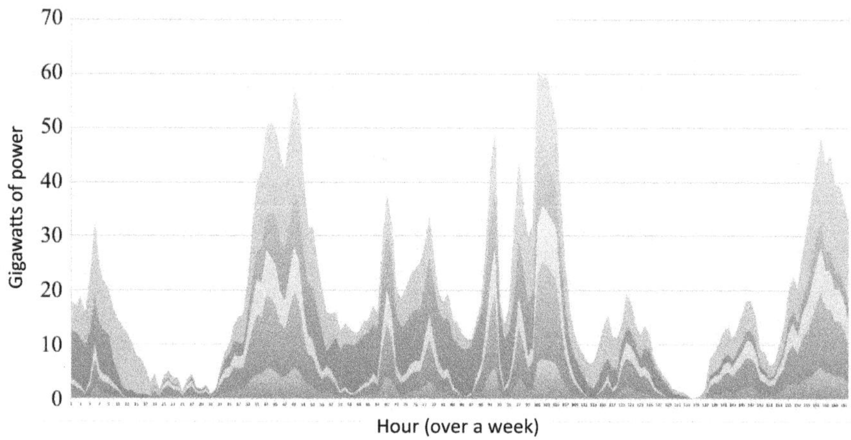

Figure 8.1: Power from five levels of wind generating capacity.

The major difficulty with renewables is handling the variability. While one solution is to simply pump any energy from a renewable source into the grid and use the capacity of the grid to handle this variability, there has been considerable interest in using storage (in particular, battery storage) to smooth out the peaks and valleys. In addition to smoothing the variability in the renewable source, there has also been interest in using batteries to take advantage of price spikes, buying power when it is cheap (prices can even go negative) and selling it back when they are high. Exploiting the variability in power prices on the grid to buy when prices are low and sell when they are high is known as battery arbitrage.

We are going to use the configuration shown in Figure 8.2 to illustrate a number of modeling and algorithmic issues in energy storage. This problem will provide insights into virtually any inventory/storage problem, including:

- Holding cash in mutual funds - A bank has to determine how much of its investment capital to hold in cash to meet requests for redemptions, versus investing the money in loans, stocks and bonds.

- Retailers (including both online as well as bricks-and-mortar stores) have to manage inventories of the hundreds of thousands of products.
- Auto dealers have to decide how many cars to hold to meet customer demand.
- Consulting firms have to decide how many employees to keep on staff to meet the variable demand of different consulting projects.

Energy storage is a particularly rich form of inventory problem. While we are not going to consider all possible variations (which are endless), our problem will exhibit the following characteristics:

- Electricity prices on the grid can be highly volatile. In the early 2000s, typical energy prices ran around \$20-\$25 per mwh, but often spiked to over \$300, and could exceed \$1000, typically during extreme weather events.
- Energy from wind can be forecasted, although not very well. Rolling forecasts are available to update these estimates.
- Solar energy exhibits three types of variability: the highly predictable process of the diurnal cycle of the sun rising and setting, the presence of very sunny or very cloudy days (these can typically be predicted a day or more in advance), and the variability of spot clouds that are difficult to predict even an hour into the future, but which can create severe power surges on the grid.
- The demand for energy is variable, but relatively predictable, since it primarily depends on temperature (and, to a lesser extent, humidity).
- Energy may be bought from or sold to the grid at current grid prices. Similarly energy from the renewables source may be used to satisfy the current load (demand for power), stored, or sold back to the grid (depending on the configuration).

- There is a roughly 5 to 10 percent loss for the conversion of power from AC (as it arrives over the grid) to DC (required to store power in the battery).

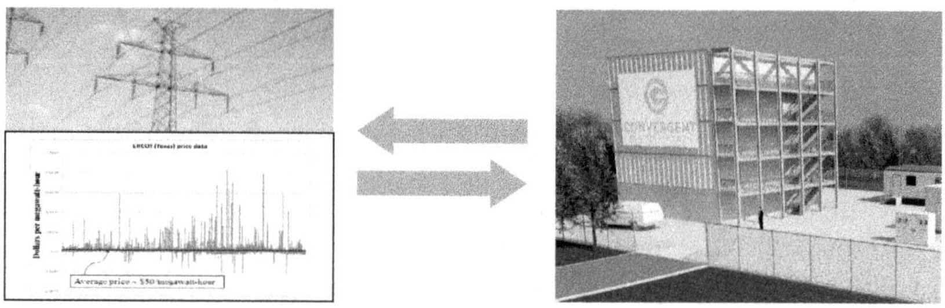

Figure 8.2: Grid-to-storage system for power stabilization and battery arbitrage.

8.2 Basic model

We will use a sequence of variations of this problem to illustrate different modeling issues, beginning with a basic system of using a battery to buy from and sell to the grid to take advantage of price volatility. For this application, we will step forward in 5-minute time increments, since this is the frequency with which prices are updated on the grid (this time increment varies depending on the grid operator).

8.2.1 State variables

For our basic model, we only need to keep track of two variables:

R_t = The amount of energy (measured in megawatt-hours, or mwh) stored in the battery at time t.

p_t = The price of energy on the grid.

Our state variable is then

$$S_t = (R_t, p_t).$$

The state variable quickly becomes more complex as we add different elements to the model. For example, the state variable to represent prices depends on how we model the price process, as described in the transition function (see below).

8.2.2 Decision variables

Our only decision is whether to buy from or sell to the grid:

$$x_t = \text{The amount of power bought from } (x_t > 0) \text{ or sold to } (x_t < 0) \text{ the grid.}$$

When we transfer energy into or out of the battery, we are going to assume that we only get a fraction η in the transfer, implying a loss of $1-\eta$. For simplicity we are going to assume that this loss is the same regardless of whether we are charging or discharging the battery.

The decision is limited by the capacity of the battery, which means we have to observe the constraints

$$\begin{aligned} x_t &\leq \frac{1}{\eta}(R^{max} - R_t), \\ x_t &\geq -\frac{1}{\eta}R_t, \end{aligned}$$

where the first constraint applies when we are buying from the grid ($x_t > 0$) while the second constraint applies when we are selling to the grid ($x_t < 0$).

As always, we assume that decisions are made with a policy $X^\pi(S_t)$, to be determined below.

8.2.3 Exogenous information

In our basic model, the only exogenous information is the change in prices. We may assume that the price in each time period is revealed, without any model to predict the price based on past prices. In this case, our exogenous information W_t would be

$$W_{t+1} = p_{t+1}.$$

Alternatively, we can assume that we observe the change in price $\hat{p}_t = p_t - p_{t-1}$, in which case we would write

$$W_{t+1} = \hat{p}_{t+1}.$$

8.2.4 Transition function

The evolution of the state variables are given by

$$R_{t+1} = R_t + \eta x_t, \tag{8.1}$$

$$p_{t+1} = p_t + \hat{p}_{t+1}. \tag{8.2}$$

This style of modeling the price process by "observing" the change in price helps us when writing the transition function. In practice, we would typically be observing p_{t+1} directly (rather than the change), in which case there is no need for an explicit transition equation. These two equations make up our transition function $S_{t+1} = S^M(S_t, x_t, W_{t+1})$.

Later, we are going to find it useful to model the post-decision state S_t^x, which is the state just after we make a decision x_t, but before any new information arrives. The post-decision resource state is

$$R_t^x = R_t + \eta x_t.$$

Because the storage variable evolves deterministically, the transition to the next pre-decision state is just

$$R_{t+1} = R_t^x.$$

The price p_t, on the other hand, is not affected by the decision, so the post-decision price would just be

$$p_t^x = p_t.$$

This means the post-decision state is

$$S_t^x = (R_t^x, p_t).$$

8.2.5 Objective function

In any period, the amount of money we make or lose is given by

$$C(S_t, x_t) = -p_t x_t.$$

Our objective function is then the canonical problem given by

$$\max_\pi \mathbb{E} \sum_{t=0}^{T} -p_t X^\pi(S_t),$$

where $S_{t+1} = S^M(S_t, X^\pi(S_t), W_{t+1})$ is given by Equations (8.1) and (8.2). We then also need to specify the initial state S_0 (that is, R_0 and p_0) and have a method for generating $W_1, W_2, \ldots,$, which we describe next.

8.3 Modeling uncertainty

In our asset selling problem in Chapter 2, we assumed that we could model prices according to

$$p_{t+1} = p_t + \varepsilon_{t+1},$$

where we then assumed that ε_{t+1} was normally distributed with mean 0 and some known variance. Figure 8.3 shows grid prices, known as "locational marginal prices" (or LMPs in the terminology of the power community) over a year, which illustrates the tremendous volatility that prices on the grid exhibit. This volatility arises because there are surges in load (or losses in power) that can create short term shortages. Since demand is inelastic (the grid is supposed to meet 100 percent of the load), prices can jump by a factor of 20 to 50 for short periods (prices are updated in 5-minute increments).

There are several methods for modeling electricity prices. Below we are going to describe four that have been used for this problem.

8.3.1 Time series models

The time series literature is quite rich, so we are just going to illustrate a basic model which represents the price p_{t+1} as a function of the recent history of prices. For illustration, we are going to use the last three time periods, which means that we would write our model as

$$\begin{aligned} p_{t+1} &= \bar{\theta}_{t0} p_t + \bar{\theta}_{t1} p_{t-1} + \bar{\theta}_{t2} p_{t-2} + \varepsilon_{t+1}, \\ &= \bar{\theta}_t^T \phi_t + \varepsilon_{t+1}, \end{aligned} \tag{8.3}$$

where

$$\phi_t = \begin{pmatrix} p_t \\ p_{t-1} \\ p_{t-2} \end{pmatrix}$$

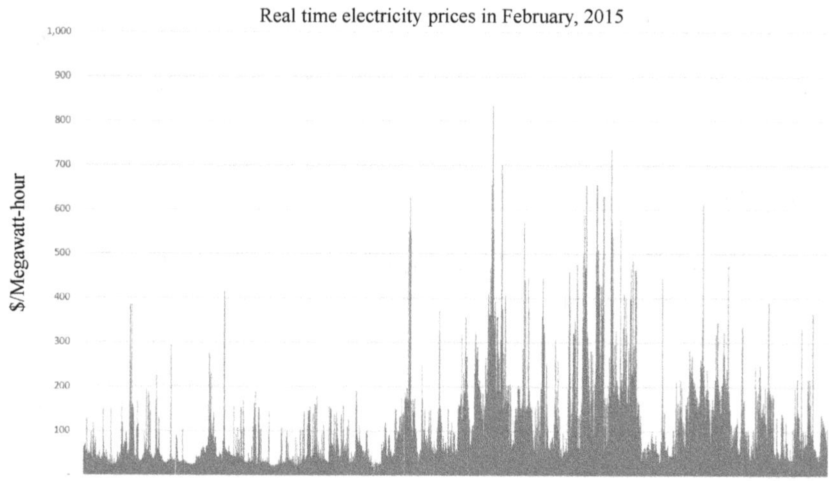

Figure 8.3: Locational marginal prices for the PJM grid (in 5-minute intervals) for 2010.

is our vector of prices. We assume that the noise $\varepsilon \sim N(0, \sigma_\epsilon^2)$ for a given σ_ϵ^2.

The vector of coefficients $\bar{\theta}_t = (\bar{\theta}_{t0}, \bar{\theta}_{t1}, \bar{\theta}_{t2})^T$ can be estimated recursively. Assume that we start with an initial estimate $\bar{\theta}_0$ of the vector of coefficients. We are also going to need a three-by-three matrix M_0 that for now we can assume is a scaled identity matrix (we provide a better idea below).

The basic updating equation for $\bar{\theta}_t$ is given by

$$\bar{\theta}_{t+1} = \bar{\theta}_t - H_t \phi_t \varepsilon_{t+1}, \tag{8.4}$$

The error $\hat{\varepsilon}_t$ is computed using

$$\varepsilon_{t+1} = \bar{\theta}_t^T \phi_t - p_{t+1}. \tag{8.5}$$

The three-by-three matrix H_t is computed using

$$H_t = \frac{1}{\gamma_t} M_t, \tag{8.6}$$

where the matrix M_t is computed recursively using

$$M_t = M_{t-1} - \frac{1}{\gamma_t}(M_{t-1}\phi_t(\phi_t)^T M_{t-1}). \tag{8.7}$$

The variable γ_t is a scalar computed using

$$\gamma_t = 1 + (\phi_t)^T M_{t-1} \phi_t. \tag{8.8}$$

These equations need initial estimates for $\bar{\theta}_0$ and M_0. One way to do this is to collect some initial data and then solve a static estimation problem. Assume you observe K prices. Let Y_0 be a K-element column factor of the observed prices $p_3, p_4, \ldots, p_{K+3-1}$ (we have to start with the third price because of our need for the trailing three prices in our model).

Then let X_0 be a matrix with K rows, where each row consists of p_k, p_{k-1}, p_{k-2}. Our best estimate of $\bar{\theta}$ is given by the normal equations

$$\bar{\theta}_0 = [(X_0)^T X_0]^{-1} (X_0)^T Y_0. \tag{8.9}$$

Finally let $M_0 = [(X_0)^T X_0]^{-1}$, which shows that the matrix M_t is the time t estimate of $[(X_t)^T X_t]^{-1}$.

There are entire families of time series models that capture the relationship of variables over time. If we were to just apply these methods directly to price data, the results would be quite poor. First, the prices are not normally distributed. Second, while prices may go negative, this is fairly rare. However, a direct application of this model would be very likely to produce negative prices if the variance σ_ϵ^2 was calibrated to the high-noise of this type of data. Finally, the behavior of the jumps in prices over time would not be realistic.

8.3.2 Jump diffusion

A major criticism of the linear model above is that it does a poor job of capturing the large spikes that are familiar in the study of electricity prices. A simple idea for overcoming this limitation is to use what is known as a *jump diffusion model*, where we add another noise term to Equation (8.3) giving us

$$p_{t+1} = \bar{\theta}_{t0} p_t + \bar{\theta}_{t1} p_{t-1} + \bar{\theta}_{t2} p_{t-2} + \varepsilon_{t+1} + \mathbb{I}_t \varepsilon_{t+1}^J. \tag{8.10}$$

Here, the indicator variable $\mathbb{I}_t = 1$ with some probability p^{jump}, and the noise ε_{t+1}^J is normally distributed with mean μ^{jump} (which is typically much larger than zero) and variance $(\sigma^{jump})^2$ which is quite large.

We have to estimate the jump probability p^{jump}, and the mean and variance $(\mu^{jump}, (\sigma^{jump})^2)$. This is done by starting with a basic model where $p^{jump} = 0$. We use this basic model to estimate σ_ϵ^2. We then choose some tolerance such as three standard deviations (that is, $3\sigma_\epsilon$), and any observations outside of this range are due to a different source of noise. Let p^{jump} be the fraction of time periods where these observations occur. Then, compute the mean and standard deviation of these observations to get $(\mu^{jump}, (\sigma^{jump})^2)$.

We do not stop here. After taking these extreme variations from the data, we should re-fit our linear model without these observations. Standard practice is to repeat this process several times until these estimates stop changing.

Jump diffusion models do a better job of replicating the tails, but it still depends on the tail behavior of the normal distribution. A better fit can be obtained by recognizing that the variance of the noise depends on the temperature, and particularly on extreme temperatures. We might group temperatures into three ranges: below freezing, above 90 degrees F, and in-between these two values. Introducing temperature dependence, while adding another variable to the set of state variables, does introduce an additional degree of complexity (the effect of this depends on the class of policy).

8.3.3 Empirical distribution

While it may be possible to fit other parametric distributions, a powerful strategy is to numerically compute the cumulative distribution from the data, creating what is often called an *empirical distribution.* To compute this, we simply sort the prices from smallest to largest. Denote this ordered sequence by $\tilde{p}_t$, where $\tilde{p}_{t-1} \leq \tilde{p}_t$. Let $T = 105,210$ which is the number of 5-minute time periods in a year. The percentage of time periods with a price less than $\tilde{p}_t$ is then t/T. We can create a cumulative distribution using

$$F_P(\tilde{p}_t) = \frac{t}{T},$$

which is illustrated in Figure 8.4.

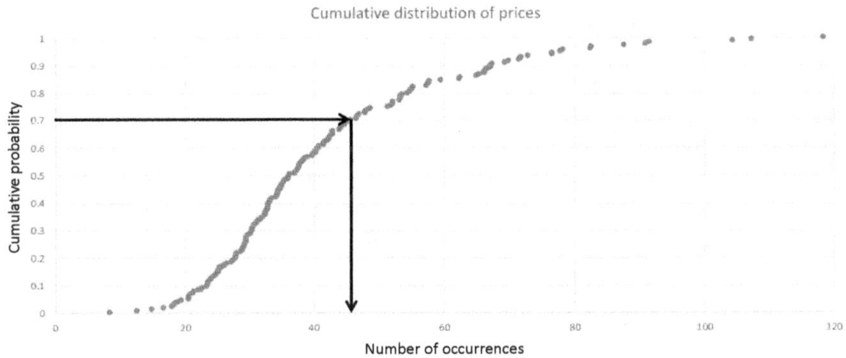

Figure 8.4: Empirical distribution of prices.

We can create a continuous distribution $F_P(p)$ for any p by finding the largest $\tilde{p}_t < p$ and setting $F_P(p)$ equal to this value, creating a step function. The function $F_P(p)$ is sometimes called a quantile distribution, and is a form of *nonparametric* distribution, since we are not fitting the distribution to any known parametric form. The good news is that it will perfectly match the data, which means that we will accurately represent the extreme tails that occur with electricity prices. The downside is that we require a good dataset to create these distributions, and we have to retain the dataset to compute the distribution, rather than just storing a small number of parameters as we would if we fit a parametric model for the distribution.

We can sample from this distribution by generating a random variable U that is uniformly distributed between 0 and 1. Let's say we generate $U = 0.70$. Then we want to find the price $p^{.70}$ that corresponds to $F_P(p^{.70}) = 0.70$, as illustrated in Figure 8.4. We write this mathematically by defining the inverse function $F_P^{-1}(u)$ which returns the price p that produces $F_P(p) = u$. We can sample repeatedly from our price distribution by just sampling the uniform random variable U and then observing a price $p = F_P^{-1}(U)$.

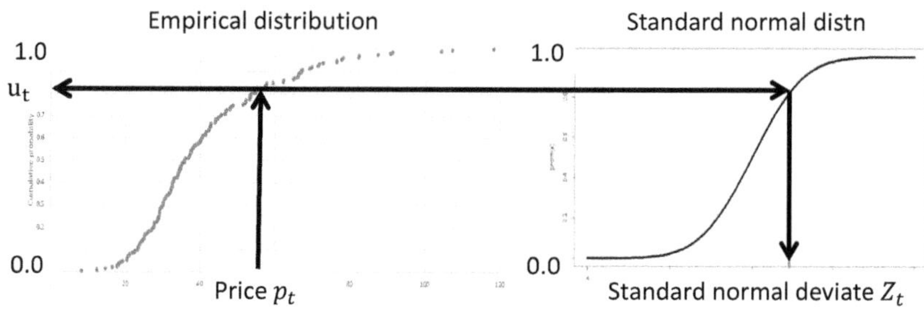

Figure 8.5: Transforming an empirical distribution to a normal distribution (and back).

8.3.4 Hybrid time-series with transformed data

A powerful strategy is to combine the use of empirical distributions with classical time series methods. We start by fitting an empirical distribution to the price data, giving us the cumulative distribution $F_P(p)$. Now, let p_t be a price and compute $u_t = F_P(p_t)$, where $0 \leq u_t \leq 1$. Next, let $\Phi(z)$ be the cumulative distribution of a standard normal random variable $Z \sim N(0, 1)$, and let $\Phi^{-1}(u)$ be its inverse. Next let $z_t = \Phi^{-1}(u_t)$. The process of mapping $p_t \to u_t \to z_t$ is illustrated in Figure 8.5.

We can use this method to transform the highly non-normal prices p_t to the sequence z_t of values that are normally distributed with mean 0 and variance 1. We can then perform any time series modeling on the sequence z_t. After this, any estimates coming from this normalized model can be transformed back to prices by tracing the path in Figure 8.5 in the reverse direction: $u_t = \Phi(z_t)$, and then $p_t = F_P^{-1}(u_t)$.

This strategy is very effective when dealing with data that is not normally distributed, and performs much better than the jump diffusion model, which is popular in finance.

8.4 Designing policies

We are going to illustrate solving this problem using two classes of policies, plus a hybrid:

Policy search - We are going to use a simple buy-low, sell-high parameterized policy. This belongs to the PFA class of policies (policy function approximations).

Lookahead policy - We will use Bellman's equation to produce a value function approximation that approximates the impact of a decision now on the future. This belongs to the VFA class of policies (policies based on value function approximations).

Hybrid policy - Finally, we are going to introduce a tunable class of policy that begins with a policy based on value functions, and then switches to policy search to further tune the policy. This will start as a VFA policy, but then transition to a parametric cost function approximation (CFA) when we use policy search to tune the parameters of what started as the value function approximation.

Policies based on Bellman's equation require computing (or approximating) the value $V_{t+1}(S_{t+1})$ which results from being in a state S_t, taking a decision x_t, and then observing random exogenous information W_{t+1}. We first saw these methods in the context of shortest path problems. One significant difference now is that the state S_{t+1} is random given S_t and x_t (in the shortest path problem, only the cost $\hat{c}_t$ was random). Also, our state variable now has two continuous dimensions, rather than just the discrete node.

We first describe the buy-low, sell-high policy, and then introduce three methods based on approximating Bellman's equation:

- Classical backward dynamic programming, which produces an optimal policy.
- Backward approximate dynamic programming.
- Forward approximate dynamic programming.

We finish with a description of a hybrid policy that combines value function approximations from Bellman's equation with a form of policy search.

8.4.1 Buy-low, sell-high

A buy-low, sell-high policy works on the simple principle of charging the battery when the price falls below a lower limit, and selling when the price goes above an upper limit. The policy can be written as

$$X^{low-high}(S_t|\theta) = \begin{cases} \text{-1} & \text{if } p_t \leq \theta^{buy}, \\ 0 & \text{if } \theta^{buy} < p_t < \theta^{sell}, \\ +1 & \text{if } p_t \geq \theta^{sell}. \end{cases} \tag{8.11}$$

We now have to tune $\theta = (\theta^{buy}, \theta^{sell})$. We evaluate our policy by following a sample path of prices $p_t(\omega)$ (or we may be observing the changes in prices $\hat{p}(\omega)$). Assuming we are generating sample paths from a mathematical model, we can generate sample paths $\omega^1, \ldots, \omega^N$. We can then simulate the performance of the policy over each sample path and take an average using

$$\overline{F}^{low-high} = \frac{1}{N}\sum_{n=1}^{N} C(S_t(\omega^n), X^{low-high}(S_t(\omega^n)|\theta)).$$

Tuning θ requires solving the problem

$$\max_\theta \overline{F}^{low-high}(\theta). \tag{8.12}$$

Since θ has only two dimensions, one strategy is to do a full grid search by discretizing each dimension, and then searching over all possible values of the two dimensions. A common discretization is to divide a region into 5-percent increments. Including the boundaries, this means we have to represent 21 values of each parameter, creating a grid of size 441 points, which is manageable (although not trivial) for most problems.

We note that a brute-force grid search only works if we run enough simulations N so that the variance in the estimate $\overline{F}^\pi(\theta)$ is relatively small. However, we have methods for doing the search for θ even with noisy estimates of the performance of the policy as presented in Sections 7.5 and 7.6.

8.4.2 Backward dynamic programming

Backward dynamic programming involves directly solving Bellman's equation

$$V_t(s_t) = \max_{x_t} \left(C_t(s, x_t) + \mathbb{E}\{V_{t+1}(S_{t+1})|S_t, x_t\} \right), \quad (8.13)$$

where $S_{t+1} = S^M(s_t, x_t, W_{t+1})$, and where the expectation is over the random variable W_{t+1}. Assume that W_{t+1} is discrete, taking values in $\mathcal{W} = \{w_1, w_2, \ldots, W_M\}$, and represent the probability distribution using

$$f^W(w|s_t, x_t) = Prob[W_{t+1} = w|s_t, x_t].$$

We write the distribution as depending on the state s_t and decision x_t, but this depends on the problem. For example, we might reasonably assume that the change in the price $p_{t+1} - p_t$ depends on the current price p_t (if prices are very high they are more likely to drop), which would be a reason to condition on s_t. We might even need the dependence on x_t if purchasing a large quantity of electricity from the grid drives up prices.

We can then rewrite Equation (8.13) as

$$V_t(s_t) = \max_{x_t} \left(C_t(s_t, x_t) + \sum_{w \in \mathcal{W}} f^W(w|s_t, x_t) V_{t+1}(S^M(s_t, x_t, w)) \right).$$

A vanilla implementation of backward dynamic programming is given in Figure 8.6 which exhibits four loops:

1) The loop stepping backward in time from T to time 0.

2) The loop over all possible states $s_t \in \mathcal{S}$ (more precisely, this is the set of possible values of the state S_t at time t).

3) The loop that would be required to search over all possible decisions x_t in order to solve the maximization problem.

4) The loop over all possible values of the random variable W that is captured in the summation required to compute $V_t(s)$.

Step 0. Initialization:

Initialize the terminal contribution $V_{T+1}(S_{T+1}) = 0$ for all states S_{t+1}.

Step 1. Do for $t = T, T-1, \ldots, 1, 0$:

Step 2. For all $s \in \mathcal{S}$, compute

$$V_t(s_t) \quad = \quad \max_{x_t} \left(C_t(s_t, x_t) + \sum_{w \in \mathcal{W}} f^W(w|s_t, x_t) V_{t+1}(S^M(s_t, x_t, w)) \right)$$

Figure 8.6: A backward dynamic programming algorithm.

It is useful to consider the range of values that each loop might take. For an energy problem, we might optimize a storage device in hourly increments over a day, giving us 24 time steps. If we use 5-minute time steps (some grid operators update prices every 5-minutes), then a 24 hour horizon would imply 288 time periods (multiply by seven if we want to plan over a week). If we are doing frequency regulation, then we have to make decisions every 2 seconds, which translates to 43,200 time periods over a day.

Our state variable consists of $S_t = (R_t, p_t)$, which means we have to replace the loop over all states, with nested loops over all values of R_t, and then all values of p_t. Since both are continuous, each will have to be discretized. The resource variable R_t would have to be divided into increments based on how much we might charge or discharge in a single time increment. We then have to discretize the grid price p_t. Grid prices can go as low as -$100, and as high as $10,000 (in extreme cases). A reasonable strategy might be to construct an empirical distribution, and then represent prices corresponding to, say, each increment of two percent of the cumulative distribution, giving us 50 possible prices.

The number of charge-discharge decisions might be as small as three (charge, discharge or do nothing), or much larger if we can charge or discharge at different rates.

Finally, the probability distribution $f^W(w)$ would be the distribution of the random changes in prices, $\hat{p}_{t+1}$. Again, we recommend constructing an empirical distribution of the changes in $\hat{p}_{t+1}$ and then discretizing the cumulative distribution into increments of, say, two percent.

If we have a two-dimensional state variable (as is the case with our basic model), then we already have five loops (time, the two state variables, the max operator over x, and then the summation over outcomes of W). This can get expensive, and we have just started. Now imagine that we are using the time series model in Equation (8.3), where we now have to keep track of the prices (p_t, p_{t-1}, p_{t-2}). In this case, our state variable would be

$$S_t = (R_t, p_t, p_{t-1}, p_{t-2}).$$

In this case, we would now have seven nested loops. While the complexity depends on the discretization of the continuous variables, running the algorithm in Figure 8.6 could easily require a year (or more).

8.4.3 Backward approximate dynamic programming

A powerful algorithmic strategy is known as "backward approximate dynamic programming." This approach progresses exactly as we just did in Figure 8.6, with one difference. Instead of looping over all the states, we choose a random sample $\hat{S}$. We then compute the value of being in state $s \in \hat{S}$ just as we originally did, and compute the corresponding value $\hat{v}$. Assume we repeat this N times, and acquire a dataset $(\hat{s}^n, \hat{v}^n), n = 1, \ldots, N$. We then use this to fit a linear approximation of the form

$$\overline{V}(s) = \theta_0 + \theta_1\phi_1(s) + \theta_2\phi_2(s) + \ldots + \theta_F\phi_F(s), \tag{8.14}$$

where $\phi_f(s), f = 1, \ldots, F$ is a set of appropriately chosen features. Examples of features might be

$$\begin{aligned}
\phi_1(s) &= R_t, \\
\phi_2(s) &= R_t^2, \\
\phi_3(s) &= p_t, \\
\phi_4(s) &= p_t^2, \\
\phi_5(s) &= p_{t-1}, \\
\phi_6(s) &= p_{t-2}, \\
\phi_7(s) &= R_t p_t.
\end{aligned}$$

Note that with a constant term θ_0, this model has only eight coefficients to be estimated. Sampling a few hundred states should be more than enough to get a good statistical approximation. This methodology is relatively insensitive to the number of state variables, and of course there is no problem if any of the variables are continuous.

We have found backward ADP to work exceptionally well on a small set of problems (see Powell, 2022, Section 15.4 for a summary of commparisons of backward ADP against benchmarks), but there are no guarantees, and its performance clearly depends on choosing an effective set of features. In one application, we reduced a run time of one month for a standard backward MDP algorithm, down to 20 minutes, with a solution that was within 5 percent of the optimal (produced by the one-month run time).

8.4.4 Forward approximate dynamic programming

Forward approximate dynamic programming works in an intuitive way. Imagine that we start with a value function approximation $\overline{V}_t^{x,n-1}(S_t^x)$ around the post-decision state S_t^x that we computed from the first $n-1$ iterations of our algorithm. We first introduced the idea of post-decision states back in Section 1.4, but this is the state immediately after we make a decision, but before any new information has arrived.

Now, imagine that we are in a particular state S_t^n during the nth iteration of our algorithm, following a sample path ω^n that guides the sampling as we step forward in time. Assume we have a function $S_t^{x,n} = S^{M,x}(S_t^n, x)$ that takes us to the post-decision state. For our energy problem where $S_t^n = (R_t^n, p_t^n)$, the post-decision state would be

$$S^{x,n} = (R_t^n + x_t^n, p_t^n).$$

We then make a decision using

$$x_t^n = \arg\max_x \left(C(S_t^n, x) + \overline{V}_t^{x,n-1}(S^{x,n})\right).$$

Given S_t^n and our decision x_t^n, we then sample $W_{t+1}(\omega^n)$ which translates to the change in the prices $\hat{p}_{t+1}^n$. We then simulate our way to the next state

$$S_{t+1}^n = (R_t^n + x_t^n, p_t^n + \hat{p}_{t+1}^n(\omega)).$$

Thus, we are just simulating our way forward in time, which means we do not care how complex the state variable is. There are different strategies for then updating the value function approximation $\overline{V}_t^{n-1}$. One strategy is illustrated in Figure 8.7, which leaves the actual update in an updating function $U^V(\cdot)$ since this depends on how we are approximating the value function.

Step 0. Initialization:

Step 0a. Initialize $V_t^{\pi,0}, \quad t \in \mathcal{T}$.

Step 0b. Set $n = 1$.

Step 0c. Initialize S_0^1.

Step 1. Do for $n = 1, 2, \ldots, N$:

Step 2. Do for $m = 1, 2, \ldots, M$:

Step 3. Choose a sample path ω^m.

Step 4: Initialize $\hat{v}^m = 0$

Step 5: Do for $t = 0, 1, \ldots, T$:

Step 5a. Solve:

$$x_t^{n,m} = \arg\max_{x_t \in \mathcal{X}_t^{n,m}} \left(C_t(S_t^{n,m}, x_t) + V_t^{\pi,n-1}(S^{M,x}(S_t^{n,m}, x_t)) \right) \tag{8.15}$$

Step 5b. Compute:

$$\begin{aligned} S_t^{x,n,m} &= S^{M,x}(S_t^{n,m}, x_t^{n,m}) \\ S_{t+1}^{n,m} &= S^M(S_t^{x,n,m}, x^{n,m}, W_{t+1}(\omega^m)). \end{aligned}$$

Step 6. Do for $t = T-1, \ldots, 0$:

Step 6a. Accumulate the path cost (with $\hat{v}_T^m = 0$)

$$\hat{v}_t^m = C_t(S_t^{n,m}, x_t^m) + \hat{v}_{t+1}^m$$

Step 6b. Update approximate value of the policy starting at time t:

$$\overline{V}_{t-1}^{n,m} \leftarrow U^V(\overline{V}_{t-1}^{n,m-1}, S_{t-1}^{x,n,m}, \hat{v}_t^m) \tag{8.16}$$

where we typically use $\alpha_{m-1} = 1/m$.

Step 7. Update the policy value function

$$V_t^{\pi,n}(S_t^x) = \overline{V}_t^{n,M}(S_t^x) \quad \forall t = 0, 1, \ldots, T$$

Step 8. Return the value functions $(V_t^{\pi,N})_{t=1}^T$.

Figure 8.7: Forward approximate dynamic programming.

Forward approximate dynamic programming is attractive because it scales to high dimensional problems. At no time are we looping over all possible states or outcomes. In fact, we can even handle high dimensional decisions x if we approximate the value function appropriately so that we can take advantage of powerful algorithms. However, forward ADP (as with backward ADP) possesses little in the way of performance guarantees.

8.4.5 A hybrid policy search-VFA policy

Whatever way we choose to approximate the value function, our policy is given by

$$X^{VFA}(S_t) = \arg\max_x \left(C(S_t, x) + \overline{V}_t^x(S_t^x)\right).$$

We simulate the policy forward, where we let ω^n represent a sample path of the exogenous information (that is, the set of changes in prices). It is frequently the case that we are going to test our policy on historical data, in which case there is only a single sample path. However, if we have developed a mathematical model of the uncertain prices, we can sample a series of sample paths that we designate by $\omega^1, \ldots, \omega^N$, and compute the average value of the policy using

$$\overline{F}^{VFA} = \frac{1}{N}\sum_{n=1}^{N}\sum_{t=0}^{T} C(S_t(\omega^n), X^{VFA}(S_t(\omega^n))).$$

When we are building a value function approximation (which belongs in the lookahead class of policies), we typically no longer have a step where we tune the policy. However, this does not mean that we cannot try. Assume that our value function is given by the linear model in Equation (8.14). We can now write our policy using

$$X^{VFA}(S_t|\theta) = \arg\max_x \left(C(S_t, x) + \sum_{f=1}^{F} \theta_f \phi_f(S_t)\right).$$

It makes sense to use one of our backward or forward ADP algorithms to get an initial estimate of θ, but as we noted above, there is no guarantee that the resulting policy is high quality. However, we can always make it better by using this as a starting point,

$$\overline{F}^{VFA}(\theta) = \frac{1}{N}\sum_{n=1}^{N}\sum_{t=0}^{T} C(S_t(\omega^n), X^{VFA}(S_t(\omega^n)|\theta)).$$

Now, we just have to solve the policy search problem that we might pose as

$$\max_{\theta} \overline{F}^{VFA}(\theta). \tag{8.17}$$

In our earlier examples for policy search, θ as a scalar, which makes this problem relatively easy. Now, θ is a vector which might have dozens of dimensions. We are going to return to this problem later.

8.5 What did we learn?

- We revisit a simple inventory problem from Chapter 1, but in the context of energy storage, with both physical and informational variables.
- We describe a variety of stochastic models for electricity prices to illustrate the richness of uncertainty modeling.
- We describe a range of policies: a PFA (buy-low, sell-high), a VFA policy (based on backward approximate dynamic programming) as well as forward approximate dynamic programming.
- Finally, we introduce the idea of estimating a linear model for a value function approximation using the techniques of approximate dynamic programming, and then perform direct policy search on the coefficients of the linear model. So, this is a VFA-based policy initially, and then turns into a form of CFA policy with a parameterized objective function.

8.6 Exercises

Review questions

8.1 — Given the heavy-tailed nature of electricity prices, what is wrong with using a time-series model?

8.2 — Briefly sketch how the we separate the prices that fall within normal variations (three standard deviations) from the more extreme observations.

8.3 — Using historical data to fit an empirical distribution should give us a probability distribution that matches history. What other errors might still be present in the stochastic model of prices?

8.4 — Describe in words what the transformation of the data in Section 8.3.4 is accomplishing.

8.5 — Classical backward dynamic programming quickly blows up from the curse of dimensionality. Describe in words how backward approximate dynamic programming (Section 8.4.3) overcomes the curse of dimensionality. For example, if we were to double the number of dimensions in the state variable, describe how this complicates backward ADP.

Problem solving questions

8.6 — Write out the five elements of the basic model for the energy storage problem as they are given in the text. Write out the objective function assuming that the policy is the buy-low, sell-high policy

$$X^{low-high}(S_t|\theta) = \begin{cases} \text{-1} & \text{if } p_t < \theta^{buy}, \\ 0 & \text{if } \theta^{buy} \leq p_t \leq \theta^{sell}, \\ +1 & \text{if } p_t > \theta^{sell}. \end{cases}$$

Write the objective function in terms of searching over the parameters of the policy. Also, write the expectation in the objective function using the nested form that reflects each random variable.

8.7 — In the section on modeling uncertainty, we introduce a time series model for prices given by

$$p_{t+1} = \bar{\theta}_{t0} p_t + \bar{\theta}_{t1} p_{t-1} + \bar{\theta}_{t2} p_{t-2} + \varepsilon_{t+1}.$$

The book describes the updating equations for the coefficient vector $\bar{\theta}_t = (\bar{\theta}_{t0}, \bar{\theta}_{t1}, \bar{\theta}_{t2})$. Remembering that the state S_t consists of *all* the information needed to model the system from time t onward, give the updated state variable and transition function to handle this price process.

Programming questions

These exercises use the Python module *EnergyStorage_I*.[1]

8.8 — Using the python module run a grid search for the parameter vector $\theta = (\theta^{buy}, \theta^{sell})$ by varying θ^{sell} over the range from 1.0 to 100.0 in increments of \$1 for prices, and varying θ^{buy} over the range from 1.0 to θ^{sell}, also in increments of \$1. The prices will be actual historical hourly prices for an 8 day period.

8.9 — Solve for an optimal policy by using the backward dynamic pro- gramming strategy in Section 8.4.2 (the algorithm has already been implemented in the python module). Assume that the price process evolves according to

$$p_{t+1} = p_t + \varepsilon_{t+1},$$

where ε_{t+1} follows a empirical distribution based on the price differences of the actual historical prices.

a) Run the algorithm where prices are discretized in increments of \$1, then \$0.50 and finally \$0.25. Compute the size of the state space for each of the three levels of discretization, and plot the run times against the size of the state space.

b) Using the optimal value function for the discretization of \$1 and compare the performance against the best buy-sell policy you found in part (a).

---

[1]See http://tinyurl.com/sdagithub

8.10 — Download the spreadsheet "Chapter8_electricity_prices".[2] Use the data in the tab "electricity prices" for the following questions:

a) Build an empirical cumulative distribution $F_P(p) = Prob[P \leq p]$ where P is a randomly chosen price for a particular hour over the one-week period in the dataset.

b) Let $F_P^{-1}(u)$ be the inverse cumulative distribution, where u is between 0 and 1. Find the price $p(u) = F_P^{-1}(u)$ corresponding to $u = 0, 0.1, 0.2, \ldots, 0.9, 1.0$. Giving each of these prices a probability of 1/11, find the cumulative distribution, and compare it to the cumulative distribution that you created in part (a). Do these appear to match?

8.11 — Using the spreadsheet "Chapter8_electricity_prices," fit a mean reversion model of the form

$$p_{t+1} = p_t + \beta(\bar{\mu}_t - p_t) + \varepsilon_{t+1} \tag{8.18}$$

where

$$\bar{\mu}_t = (1 - \alpha)\bar{\mu}_{t-1} + \alpha p_t.$$

Assume $\alpha = 0.15$. Find β that minimizes

$$G(\beta) = \sum_{t=0}^{T} \left(p_{t+1} - (p_t + \beta(\bar{\mu}^t - p_t))\right)^2. \tag{8.19}$$

a) Fit the mean reversion model by doing a simple one-dimensional search (e.g. try values between 0 and 1 in increments of 0.1).

b) Compute the standard deviation σ of ε from your sample (we assume the mean is 0). Note that we assume there is a single, constant standard deviation, although we allow the mean $\bar{\mu}_t$ to vary over time.

c) Using the value of β you found in (a), generate 10 sample paths using Equation (8.18) by sampling $t+1$ from a normal distribution with mean 0 and standard deviation σ. Plot the sample paths in a graph, and compare the behavior of your sample paths to the historical prices. Do they seem similar?

[2] https://tinyurl.com/sdamodelingsupplements

8.12 — Now you are going to fit a jump diffusion model which is given by

$$p_{t+1} = p_t + \beta(\bar{\mu}_t - p_t) + \varepsilon_{t+1} + J_{t+1}\varepsilon_{t+1},$$

where $J_{t+1} = 1$ with some jump probability (that we calculate below) and 0 otherwise, and ε^J_{t+1} is the random size of the jump when they happen.

Follow the steps below to fit the jump diffusion model and compare the results to history.

a) Using the value of β from Exercise 8.11, pass over the data and identify all the data points that fall outside of the range $[\bar{\mu}_t \pm 3\sigma]$.

b) Using the same value of β you found in Exercise 8.11, run the mean reversion over the data a second time, but this time only include the data points that were not excluded in part (a). Find the new mean and standard deviation of σ on the data that was not excluded.

c) Repeat (b) one more time on the datapoints that were retained, again excluding data points outside the $\pm 3\sigma$ range.

d) Compute the probability of a jump as the fraction of points that were excluded by the time you finish part (c) (by now you have run the process of excluding datapoints twice). Also, compute the mean and standard deviation of the points that were excluded.

e) Now, run 10 simulations of your jump diffusion model, using the final mean and variance for the retained and excluded points, and where you sample jumps using the jump diffusion probability. Compare these simulations to history, and discuss whether the resulting price paths are more realistic than what you found in Exercise 8.11, and comparing the price paths to actual history.

8.13 — We are going to try again to get a good fit of the prices by repeating parts of Exercises 8.11 and 8.12 using transformed prices.

a) Using the cumulative distribution from Exercise 8.10, convert each of the prices to a uniformly distributed random variable using the identity $U_t = F_P(pt)$.

b) Next convert your uniformly distributed random variables U_t into normally distributed random variables with mean 0 and variance 1 using $Z_t = \Phi^{-1}(U_t)$ where $\Phi(z)$ is the cumulative distribution of the standard normal distribution, and $\Phi^{-1}(U_t)$ is the inverse of this distribution. This is captured by the norm.s.inv(p) function in Excel, which returns the Z value corresponding to a probability p (which is given by the variable U_t).

c) We now have to refit β in the mean reversion equation 8.11. This time, instead of using the price p_t, we use the normalized quantity Z_t; otherwise everything is the same (so you can just follow the process when you fit β for the raw prices).

d) Now use your model from (c) (with the new value for β) to create a sample path of Z_t values. Then, use $U_t = norm.s.dist(Z_t, 1)$ to get the probability that $Z_t \leq z$ (which is uniformly distributed between 0 and 1). Finally, map the U_t value back to a price using the cumulative distribution you found in Exercise 8.11. Plot one sample path (this is not too hard if you are proficient in Excel - otherwise the painful part is this last step).

e) Compare the behavior of the resulting sample path to the historical distribution. Note that the distribution of prices should be perfect, but the series of prices may still not look like a good fit. What mistakes might we still be making?

9

Energy Storage II

9.1 Narrative

We are now going to solve a somewhat more complex energy storage problem depicted in Figure 9.1. In contrast with our previous storage system that just bought and sold energy from the grid, we now face the problem of meeting a time-dependent load for a building using energy from a wind farm and the grid, with a single energy storage device to help smooth the different processes.

This problem is also going to exhibit another distinctive property, which is that all the exogenous processes (wind, prices, loads and temperature) are going to come from a dynamic process that varies with different types of predictability:

- Loads - The load (which is the demand for energy) follows a fairly predictable pattern that depends on time of day (a building needs to be at a particular temperature by 8am when people start showing up) as well as temperature.

- Temperature - The temperature is a reasonably predictable process that depends on time of day and season, but also reflects local weather conditions that can be forecasted with some accuracy.

- Wind - There are vendors who perform forecasting services for wind, although the forecasts are not very accurate and evolve fairly quickly even over the course of the day (see Figure 9.2).
- Prices - Prices are correlated with temperature which can be forecasted, but they require their own forecast model.

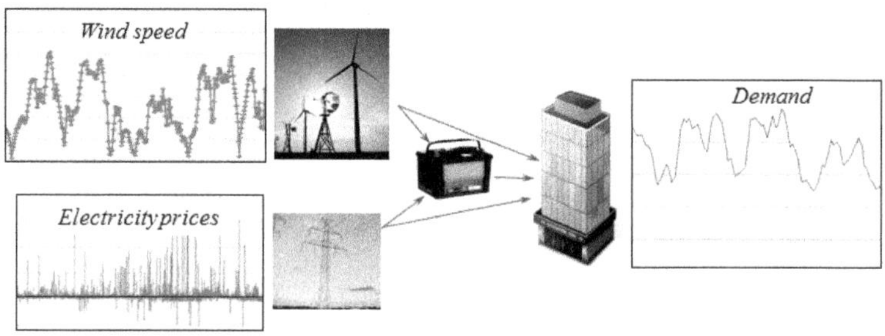

Figure 9.1: Energy system to serve a load (building) from a wind farm (with variable wind speeds) the grid (with variable prices), and a battery storage device.

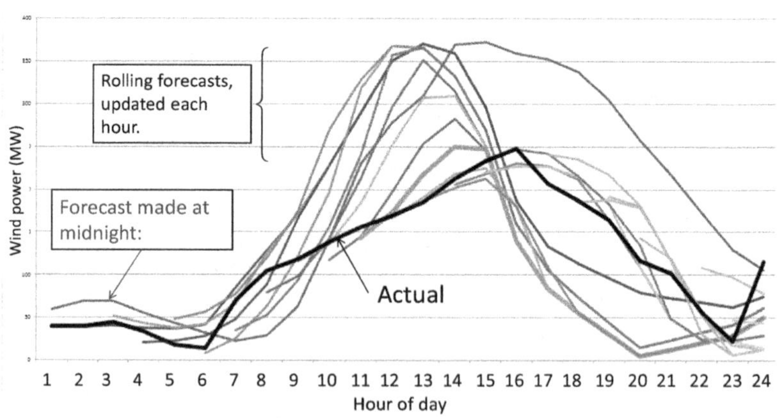

Figure 9.2: Evolution of forecasts of wind power over the course of a 24-hour period, updated each hour. The black line is the actual.

Each of these processes can be forecasted with varying degrees of accuracy. Forecasting wind power is the least accurate, and while wind can be stronger at night, highs and lows can occur at any time of the day or night. Loads are highly correlated with hour of day, largely because of human activity but also because of temperature. Note that hot afternoons can create peaks in the middle of a summer day from air conditioning, while it can actually reduce the heating load (which may be served by electrical heating) during the winter. Temperature also has a strong hour-of-day component due to the rising and setting of the sun, but there can be variations as weather fronts move through.

Our problem involves deciding how much to buy from the grid (or sell back to the grid) and how much to store at each point in time (these decisions might be made in 5-minute increments for some grid operators). We need to meet the demand for energy, but otherwise would like to maximize the revenue we make from selling energy minus the cost of buying the energy from the grid or wind farm.

9.2 Basic model

9.2.1 State variables

We start by modeling the snapshot of the system at time t which includes

$$\begin{aligned}
R_t &= \text{the amount of energy (in MWh) stored in the battery at time } t, \\
L_t &= \text{the load (demand) for energy at time } t \text{ (in MW)}, \\
\tau_t &= \text{the temperature at time } t, \\
w_t &= \text{the energy from wind at time } t \text{ (in MW)}, \\
p_t^{load} &= \text{the amount we are paid per MWh to satisfy the load to the building at time } t, \\
c_t^{grid} &= \text{the cost of purchasing power from the grid (this is the price we are paid if we sell back to the grid).}
\end{aligned}$$

Since the underlying problem is highly nonstationary, we are going to need to use forecasts, both to model the dynamics of the problem as well as to make decisions which need to anticipate what might happen in the future. We assume that we are given a rolling set of forecasts as illustrated for wind in Figure 9.2. We model the forecasts for load

(L), temperature (T), wind (W), market prices (p), and grid prices (G) using:

$f^L_{tt'}$ = Forecast of the load L_t (in MW) at time $t' > t$ given what we know at time t.

$f^\tau_{tt'}$ = Forecast of the temperature τ_t at time $t' > t$ given what we know at time t.

$f^w_{tt'}$ = Forecast of the wind power w_t (in MW) at time $t' > t$ given what we know at time t.

$f^p_{tt'}$ = Forecast of market prices p^{load}_t (in \$/MWh) at $t' > t$ given what we know at time t.

$f^G_{tt'}$ = Forecast of grid prices c^{grid}_t (in \$/MWh) at $t' > t$ given what we know at time t.

All forecasts are vectors over the horizon $t, t+1, \ldots, t+H$ where H is a specified horizon (e.g. 24 hours). We let f^X_t be the vector of forecasts for $X \in \mathcal{X} = \{L, T, W, P, G\}$.

Our state variable is then

$$S_t = (R_t, (L_t, \tau_t, w_t, p^{load}_t, c^{grid}_t), (f^L_t, f^T_t, f^w_t, f^P_t, f^G_t)).$$

Here we have grouped the controllable resource R_t (our physical state variable), the snapshot of load, temperature, wind power and price (which we might group as information variables I_t), and then the forecasts (which represent a form of belief B_t about the future).

We quickly see that we have a relatively high-dimensional state variable. If we are planning in 5-minute increments, a rolling 24-hour forecast would have 288 elements. Our five forecasts alone would represent 1,440 elements (dimensions). This is hardly record-setting, but it hints at the challenge awaiting anyone who wants to estimate the value $V_t(S_t)$ of being in state S_t.

9.2.2 Decision variables

The decision variables for our system are now

x^{wr}_t = the amount of power moved from the wind farm to the battery at time t,

$x^{w\ell}_t$ = the amount of power moved from the wind farm to the load (the building) at time t,

x_t^{gr} = the amount of power moved from the grid to the battery at time t,

x_t^{rg} = the amount of power moved from the battery to the grid at time t,

$x_t^{g\ell}$ = the amount of power moved from the grid to the load at time t,

$x_t^{r\ell}$ = the amount of power moved from the battery to the load at time t,

x_t^{loss} = uncovered load (known as "load shedding").

These variables have to be determined subject to the constraints

$$x_t^{w\ell} + x_t^{g\ell} + \eta x_t^{r\ell} + x_t^{loss} = L_t, \tag{9.1}$$

$$x_t^{r\ell} + x_t^{rg} \leq R_t, \tag{9.2}$$

$$x_t^{wr} + x_t^{gr} \leq \frac{1}{\eta}(R^{max} - R_t), \tag{9.3}$$

$$x_t^{w\ell} + x_t^{wr} \leq w_t, \tag{9.4}$$

$$x_t^{wr} + x_t^{gr} \leq u^{charge}, \tag{9.5}$$

$$x_t^{r\ell} + x_t^{rg} \leq u^{discharge}, \tag{9.6}$$

$$x_t^{wr}, x_t^{w\ell}, x_t^{gr}, x_t^{rg}, x^{g\ell}, x_t^{r\ell} \geq 0. \tag{9.7}$$

Equation (9.1) limits the power to serve the load (the building) to the amount of the load (we cannot overload the building). The variable x^{loss} captures the amount by which we have not covered the load. The constraint captures conversion losses for energy taken from the battery (similar losses might apply to the energy from the wind farm, since this might be DC power that has to be converted to AC). Equation (9.2) says that we cannot move more power out of our battery storage than is in the battery. Equation (9.3) then limits how much we can move into the battery to the amount of available capacity, adjusted by the conversion losses. Equation (9.4) limits the amount from the wind farm to what the wind farm is generating at that point in time. Equations (9.5) - (9.6) limit the flows into and out of the battery to charge and discharge rates.

As always, we assume that decisions are made with a policy $X^\pi(S_t)$, to be determined below. For this problem, the policy has to return a five-dimensional vector.

9.2.3 Exogenous information

Following the lead of the first energy model, we are going to represent exogenous information in the form of changes. We again let f_t^X be a vector of forecasts for each information process X (such as load, temperature, wind, market prices and grid prices), where f_{tt}^X is the actual value at time t. Let

$$
\begin{aligned}
\hat{f}^X_{t+1,t'} &= \text{the change in the forecast for time } t' \text{ between } t \text{ and } \\
&\quad t+1, \\
&= f^X_{t+1,t'} - f^X_{tt'}, \ t' = t, t+1, \ldots, t+H.
\end{aligned}
$$

The exogenous changes $\hat{f}^X_{t+1,t'}$ are correlated across the time periods t'. If this were not the case, then the forecasts when plotted over the horizon $t' = t, \ldots, t+H$ would no longer exhibit the smoothness that we see in the wind forecasts in Figure 9.2. We return to the issue of modeling the uncertainty in forecasts in Section 9.3 on uncertainty modeling.

9.2.4 Transition function

The evolution of the resource state variable is given by

$$
R_{t+1} = R_t + \eta(x_t^{wr} + x_t^{gr} - x_t^{rg} - x_t^{r\ell}). \tag{9.8}
$$

Each of the variables L_t, τ_t, w_t, p_t^{load}, and c_t^{grid} evolve using the forecasts. For example, we would write the evolution of the load L_t using

$$
L_{t+1} = f^L_{t,t+1} + \varepsilon^L_{t+1}, \tag{9.9}
$$

where $\varepsilon^L_{t+1} \sim N(0, \sigma_L^2)$ where σ_L^2 is the variance of the error. We could create similar equations for τ_t, w_t, p_t^{load}, and c_t^{grid}.

We write the evolution of the forecasts using

$$
f^X_{t+1,t'} = f^X_{tt'} + \hat{f}^X_{t+1,t'}, \ X \in \mathcal{X}, \ t' = t+1, \ldots, t+1+H, \tag{9.10}
$$

for $X = L, \tau, w, p^{load}$ and c^{grid}. Equation (9.10) is known in the literature as the "martingale model of forecast evolution." The term "martingale" refers to our assumption that $f^X_{tt'}$ is an unbiased estimate of $f_{t+1,t'}^X$ since we assume that the random deviations $\hat{f}_{t+1,t'}^X$ are, on average, zero.

Equations (9.8), (9.9) and (9.10) (for all forecasts X) make up the transition function

$$S_{t+1} = S^M(S_t, x_t, W_{t+1}).$$

9.2.5 Objective function

Our profit function at time t is given by

$$C(S_t, x_t) = (x_t^{w\ell} + x_t^{g\ell} + \eta x_t^{r\ell})p_t^{load} - (x_t^{g\ell} + x_t^{gr})c_t^{grid},$$

where the market price p_t^{load} and grid price c_t^{grid} are contained in the state variable S_t. Our objective function is still the canonical problem given by

$$\max_{\pi} \mathbb{E} \sum_{t=0}^{T} p_t X^{\pi}(S_t),$$

As before, $S_{t+1} = S^M(S_t, X^{\pi}(S_t))$ which is given by Equations (9.8), (9.9) and (9.10).

9.3 Modeling uncertainty

We describe below two styles for modeling uncertainty over time. We first describe a method for modeling the correlations in errors in the forecasts over time using a technique sometimes called *Gaussian process regression.* This method ensures that as we move forward in time, a vector of forecasts evolves in a natural way.

Then we describe a hidden-state Markov model that we have found provides exceptionally realistic sample paths for stochastic processes. This model closely replicates error distributions, but also does a very nice job capturing *crossing times*, which is the time that the actual process (e.g. wind speed) stays above or below a benchmark such as a forecast.

This section will illustrate several powerful methods for stochastic modeling that should be in any toolbox for modeling uncertainty.

9.3.1 Gaussian process regression for forecast errors

Gaussian process regression (GPR for short) is a simple method for generating correlated sequences of normally distributed random variables. GPR is particularly useful when trying to estimate a continuous surface, where if one point in the surface is higher than expected, then it means that nearby points will also be higher than expected.

Let $X_{t'}$ be the actual outcome of any of our exogenous processes (prices, loads, temperature, wind energy) at time t', and let $f^X_{tt'}$ be the forecast of $X_{t'}$ made at time $t < t'$. It is common to assume some error $\varepsilon_{t'-t}$ that describes the difference between $X_{t'}$ and the forecast $f^X_{tt'}$. We would then assume some model for $\varepsilon_{t'-t}$ such as

$$\varepsilon^X_{t'-t} \quad \sim \quad N(0, (t'-t)\sigma^2_X).$$

We are going to take a somewhat different approach by assuming that the distribution in the change in a forecast $\hat{f}^X_{t+1,t'}$ is described by

$$\hat{f}^X_{t+1,t'} \quad \sim \quad N(0, \sigma^2_X).$$

We then assume that the changes in the forecasts $\hat{f}^X_{t+1,t'}$ are correlated across times t' with a covariance function

$$Cov(\hat{f}^X_{t+1,t'}, \hat{f}^X_{t+1,t''}) \quad = \quad \sigma^2_X e^{-\beta|t''-t'|}, \tag{9.11}$$

We can use this covariance function to create a covariance matrix Σ^X with element $\Sigma^X_{t't''} = \sigma^2_X e^{-\beta|t''-t'|}$. There is a simple way of creating a correlated sample of changes in forecasts using a method called Cholesky decomposition. It begins by creating what we might call the "square root" of the covariance matrix Σ^X which we store in a lower triangular matrix L. In python, using the NumPy package, we would use the python command

$$L \quad = \quad \texttt{scipy.linalg.cholesky}(\Sigma^X, lower = True).$$

We note that

$$\Sigma^X \quad = \quad L^T L,$$

which is why we think of L as the square root of Σ^X.

Next generate a sequence of independent random variables Z_τ for $\tau = 1, \ldots, H$ that are normally distributed with mean 0 and variance 1. Now let $Z = (Z_{t+1}, Z_{t+2}, \ldots, Z_{t+H})^T$ be a column vector made up of these independently distributed standard normal random variables. We can create a correlated sample of changes in the forecasts using

$$\begin{pmatrix} \hat{f}^X_{t+1,t+1} \\ \hat{f}^X_{t+1,t+2} \\ \vdots \\ \hat{f}^X_{t+1,t+H} \end{pmatrix} = \begin{pmatrix} 0 \\ 0 \\ \vdots \\ 0 \end{pmatrix} + LZ.$$

This formula gives us a sampled set of changes in forecasts $\hat{f}^X_{t+1,t+1}, \ldots, \hat{f}^X_{t+1,t+H}$ that are correlated according to our exponential decay function in Equation (9.11). The result will be an evolving set of forecasts where the variance in the errors in the forecasts grows linearly over time according to

$$Var(\varepsilon^X_{t'-t}) = (t'-t)\sigma^2_X.$$

The evolving forecasts $f^X_{t,t'}, f^X_{t+1,t'}, \ldots$ will exhibit the behavior that we saw in our evolving wind forecasts in Figure 9.2.

9.3.2 Hidden-state Markov model

A challenge when developing stochastic models in energy is capturing a property known as a *crossing time*. This is the time that an actual process (e.g. price or wind speed) is above or below some benchmark such as a forecast. Figure 9.3 illustrates an up-crossing time for a wind process.

Replicating crossing times using standard time series modeling proved unsuccessful. What did work was the development of a Markov model with a hidden state variable S^C_t which is calibrated to capture the dynamics of the process moving above or below the benchmark. The process uses the following steps:

Step 1 Comparing the actual process to the benchmark, find the times at which the actual process moves above or below the benchmark, and output a dataset that captures whether the process was above (A) or below (B) and for how long. Aggregate these periods into

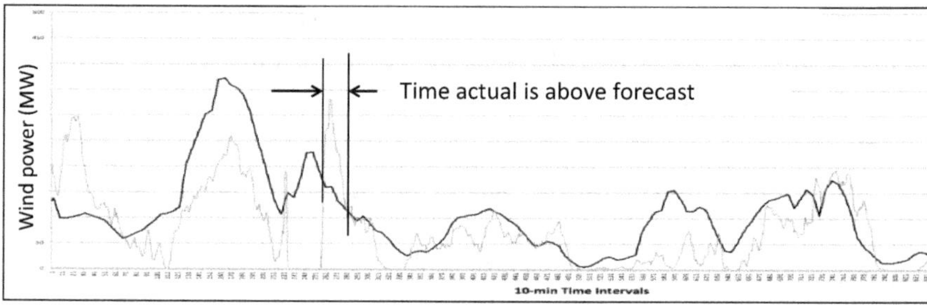

Figure 9.3: Forecasted (black) and actual wind power, illustrating a period where the actual is above the forecast. The length of this period of being above is called an up-crossing time.

three buckets (S/M/L) for short/medium/long, and label each segment with A or B and S/M/L, creating six states. These are called "hidden states" because, while we will know at time t if the actual process is above or below the benchmark, we will not know if the length is short, medium or long until after the process crosses the benchmark.

Step 2 Using the historical sequence of S_t^C, compute a one-step transition matrix

$$P^C[S_{t+1}^C|S_t^C] \quad = \quad \text{the probability that the crossing process takes on value } S_{t+1}^C \text{ given that it is currently in state } S_t^C.$$

Step 3 Aggregate the actual process (e.g. wind speed) into, say, five buckets based on the empirical cumulative distribution. Let W_t^g be the aggregated wind speed (a number from 1 to 5).

Step 4 From history, compute the conditional distribution of wind speed given W_t^g and S_t^C,

$$F^W[W_{t+1}|W_t^g, S_t^C] \quad = \quad \text{empirical cumulative distribution of the wind speed } W_{t+1} \text{ given } W_t^g \text{ and } S_t^C.$$

Using the one-step transition matrix $P^C[S_{t+1}^C|S_t^C]$ and the conditional cumulative distribution $F^W[W_{t+1}|W_t^g, S_t^C]$, we can now simulate our

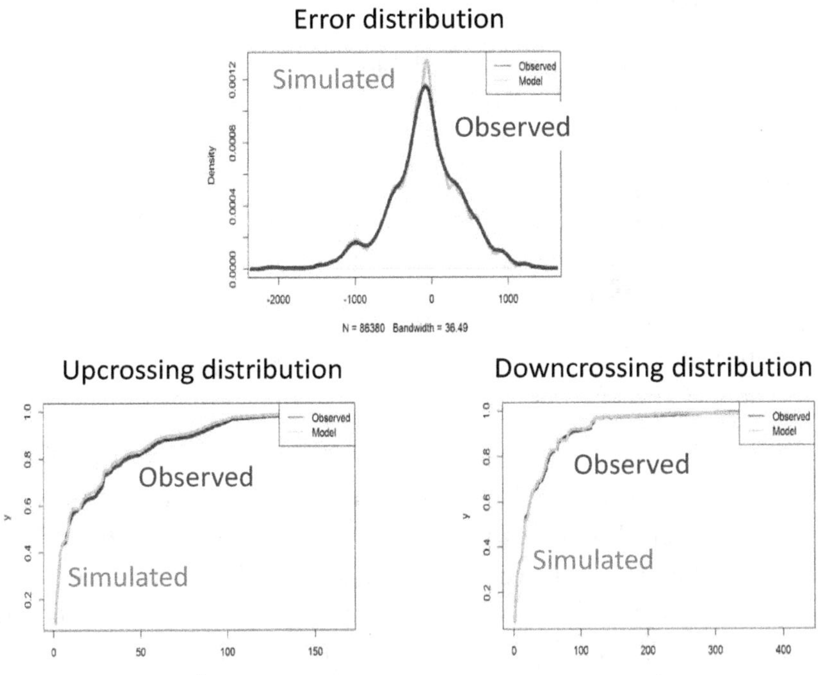

Figure 9.4: Comparison of actual vs predicted forecast error distributions (top), up-crossing time distributions (bottom left) and down-crossing time distributions (bottom right).

stochastic process by first simulating the hidden state variable S^C_{t+1} given S^C_t (note that there are only 30 of these). Then, from a wind speed W_t, we can find the aggregated wind speed W^g_t, and then sample the actual wind speed W_{t+1} from the conditional cumulative distribution $F^W[W_{t+1}|W^g_t, S^C_t]$.

This logic has been found to accurately reproduce both the error distribution (actual vs. benchmark), as well as both up-crossing and down-crossing distributions across a range of datasets modeling wind as well as grid prices. Figure 9.4 illustrates these distributions on a particular dataset.

9.4 Designing policies

The biggest complication of this problem is that the forecasts are part of the state variable, which allows us to explicitly model the rolling evolution of the forecast. The difficulty is that this makes the state variable high-dimensional.

The most common approaches for handling forecasts use a lookahead model that approximates the future by fixing the forecast. The two most popular strategies are:

- Deterministic lookahead with the forecast captured in the formulation of the lookahead.
- Stochastic lookahead with latent variables - We can use the forecast to develop a stochastic lookahead model which we then solve using classical dynamic programming. In the lookahead model, we fix the forecasts into the model, rather than modeling their evolution over time. When we ignore a variable in a model (including a lookahead model), it is called a *latent variable.*

Both of these methods use the forecast as a latent variable in that they do not model the evolution of the forecast within the lookahead model. The difficulty with the stochastic lookahead model is that it is harder to solve. If we are optimizing our energy storage problem in short-time increments (this might be 5-minutes, or even less), then this can create problems for techniques such as exact or even approximate dynamic programming.

For this reason, the most popular strategy for handling time-dependent problems with a forecast is to solve a deterministic lookahead model, just as we did for our dynamic shortest path problem. We describe such a model below, and then introduce a parameterized version that allows the deterministic model to better handle uncertainty.

9.4.1 Deterministic lookahead

We are going to use the same notational style we first introduced in Chapter 6, where we distinguish our *base model*, which is the problem we are trying to solve, from the *lookahead model* which we solve as a form of policy for solving the base model.

Recall that the canonical formulation of our base model is

$$\max_{\pi} \mathbb{E}\left\{\sum_{t=0}^{T} C(S_t, X^{\pi}(S_t))|S_0\right\},$$

where $S_{t+1} = S^M(S_t, X^{\pi}(S_t), W_{t+1})$, and where we have an exogenous information process $(S_0, W_1, W_2, \ldots, W_T)$. Note that the variables W_t may depend on the state S_t and/or decision x_t; if this is the case, then the variable W_{t+1} has to be generated on the fly after we know S_t and x_t.

We are going to create a policy by formulating a deterministic lookahead model, where all the variables are labeled with tilde's, and indexed both by the time t at which we are making our decision, and time t' which is the time variable within the lookahead model. So we would define

$$\begin{aligned}
\tilde{x}_{tt'} &= \text{the decision at time } t' \text{ in the lookahead model being generated at time } t,\\
\tilde{c}_{tt'} &= \text{the cost coefficient for } \tilde{x}_{tt'},\\
\tilde{R}_{tt'} &= \text{the energy in the battery at time } t' \text{ in the lookahead model generated at time } t.
\end{aligned}$$

Note that $x_t = \tilde{x}_{tt}$, $c_t = \tilde{c}_{tt}$ and so on.

We create our deterministic lookahead policy $X_t^{DLA}(S_t)$ as the following linear program:

$$X_t^{DLA}(S_t) = \underset{x_t, (\tilde{x}_{tt'}, t'=t+1,\ldots,t+H)}{\arg\max} C(S_t, x_t) + \sum_{t'=t+1}^{t+H} C(\tilde{S}_{tt'}, \tilde{x}_{tt'}), \quad (9.12)$$

where

$$\begin{aligned}
C(S_t, x_t) &= (x_t^{w\ell} + x_t^{g\ell} + \eta x_t^{r\ell})p_t^{load} - (x_t^{g\ell} + x_t^{gr})c_t^{grid}, &(9.13)\\
C(\tilde{S}_{tt'}, \tilde{x}_{tt'}) &= (\tilde{x}_{tt'}^{w\ell} + \tilde{x}_{tt'}^{g\ell} + \eta \tilde{x}_{tt'}^{r\ell})\tilde{p}_{tt'}^{load} - (\tilde{x}_{tt'}^{g\ell} + \tilde{x}_{tt'}^{gr})\tilde{c}_{tt'}^{grid}. &(9.14)
\end{aligned}$$

This problem has to be solved subject to the constraints (9.1)-(9.7) for x_t, and the following constraints for $\tilde{x}_{tt'}$ for all $t' = t+1, \ldots, t+H$:

$$\tilde{R}_{t,t'+1} - (R_{tt'} - x_t^{r\ell} + \eta x_{tt'}^{wr} + \eta x_{tt'}^{gr} - x_{tt'}^{rg}) = 0, \tag{9.15}$$

$$\tilde{x}_{tt'}^{w\ell} + \tilde{x}_{tt'}^{g\ell} + \eta \tilde{x}_{tt'}^{r\ell} + \tilde{x}_{tt'}^{loss} \leq f_{tt'}^{L}, \tag{9.16}$$

$$\tilde{x}_{tt'}^{r\ell} + \tilde{x}_{tt'}^{rg} \leq \tilde{R}_{tt'}, \tag{9.17}$$

$$\tilde{x}_{tt'}^{wr} + \tilde{x}_{tt'}^{gr} \leq \frac{1}{\eta}(R^{max} - \tilde{R}_{tt'}), \tag{9.18}$$

$$\tilde{x}_{tt'}^{w\ell} + \tilde{x}_{tt'}^{w\ell} \leq f_{tt'}^{W}, \tag{9.19}$$

$$\tilde{x}_{tt'}^{wr} + \tilde{x}_{tt'}^{gr} \leq u^{charge}, \tag{9.20}$$

$$\tilde{x}_{tt'}^{r\ell} + \tilde{x}_{tt'}^{rg} \leq u^{discharge}, \tag{9.21}$$

$$\tilde{x}_{tt'} \geq 0. \tag{9.22}$$

These equations mirror the ones in the base constraints (9.1)-(9.7), with the only change that we use the lookahead variables such as $\tilde{x}_{tt'}$, $\tilde{R}_{tt'}$, and forecasts such as $f_{tt'}^{W}$ instead of the actual wind W_t.

The model described by Equations (9.12) - (9.22) is a relatively simple linear program, for which packages are now available in languages such as Matlab or python.

Lookahead policies such as $X_t^{DLA}(S_t)$ are widely used in dynamic, time-varying problems such as this. They have to be solved on a rolling basis as we first illustrated in Figure 6.1 for our deterministic shortest path problem. For this reason, these are sometimes called "rolling horizon procedures" or "receding horizon procedures." There is an entire field known as "model predictive control" which is based on these lookahead policies.

For applications such as this energy storage problem, the use of a deterministic lookahead model raises the concern that we are not accounting for uncertainties. For example, we might want to store extra energy in the battery to protect ourselves from a sudden drop in wind or a surge in prices on the grid.

9.4.2 Parameterized lookahead

There is a very simple way to address the problem that our deterministic lookahead does not handle uncertainty. What we need to do is to think about how we might modify the model (or the solution) because of uncertainty. For example, we might wish to pay for extra storage *in the future* to handle unexpected variations. Of course, we cannot force the model to keep energy in storage right now when we might need it. We might also want to discount forecasts that might not be very accurate.

We can introduce these changes by replacing the constraints (9.15)-(9.22) with the following

$$
\begin{aligned}
\tilde{R}_{t,t'+1} - (\tilde{R}_{tt'} - \tilde{x}^{r\ell}_{t} + \eta\tilde{x}^{wr}_{tt'} + \eta\tilde{x}^{gr}_{tt'} - \tilde{x}^{rg}_{tt'}) &= 0, && (9.23)\\
\tilde{x}^{w\ell}_{tt'} + \tilde{x}^{g\ell}_{tt'} + \eta\tilde{x}^{r\ell}_{tt'} + \tilde{x}^{loss}_{tt'} &= \theta^{L}_{t'-t} f^{L}_{tt'}, && (9.24)\\
\tilde{x}^{r\ell}_{tt'} + \tilde{x}^{rg}_{tt'} &\leq \theta^{R}_{t'-t}\tilde{R}_{tt'}, && (9.25)\\
\tilde{x}^{wr}_{tt'} + \tilde{x}^{gr}_{tt'} &\leq \frac{1}{\eta}(R^{max} - \tilde{R}_{tt'}), && (9.26)\\
\tilde{x}^{w\ell}_{tt'} + \tilde{x}^{w\ell}_{tt'} &\leq \theta^{W}_{t'-t} f^{W}_{tt'}, && (9.27)\\
\tilde{x}^{wr}_{tt'} + \tilde{x}^{gr}_{tt'} &\leq u^{charge}, && (9.28)\\
\tilde{x}^{r\ell}_{tt'} + \tilde{x}^{rg}_{tt'} &\leq u^{discharge}, && (9.29)\\
\tilde{x}_{tt'} &\geq 0. && (9.30)
\end{aligned}
$$

Note that we have introduced parameters to modify the right hand side of constraints (9.24) and (9.27) where we have introduced coefficients $\theta^L_{t'-t}$ and $\theta^W_{t'-t}$ to modify the forecasts of load and wind, where the coefficients are indexed by how many time periods we are forecasting into the future. Then, we modified the constraint (9.25) with the idea that we may want to restrict our ability to use all the energy in storage in order to maintain a reserve. The coefficient $\theta^R_{t'-t}$, is also indexed by how many time periods we are looking into the future.

Let $X^{DLA-P}(S_t|\theta)$ represent the lookahead policy that is solved subject to the parameterized constraints (9.23) - (9.30). Once we have decided how to introduce these parameterizations (this is the art behind any parametric model), there is the problem of finding the best value for θ. This is the problem of parameter search that we addressed in Chapter 7.

9.5 What did we learn?

- In this chapter we introduce a much more complex energy storage problem with rolling forecasts.
- We introduce the "martingale model of forecast evolution" which assumes that forecasts of the future evolve over time, where the expected change in a forecast is zero (but the actual change is positive or negative).
- We then describe a hidden semi-Markov model that helps us replicate "crossing times" which capture the time that a forecast is above or below actual.
- We introduce a deterministic lookahead policy, and then a parameterized deterministic lookahead, where we introduce coefficients for each forecast (another example of a hybrid DLA/CFA).
- We show that the tuned DLA/CFA policy outperforms the pure (untuned) deterministic lookahead by around 30 percent.

9.6 Exercises

Review questions

9.1 — What is meant by the "martingale model of forecast evolution"?

9.2 — The entire forecast at time t for each quantity such as the load L_t, $f^L_{tt'}$ for $t' = t, \ldots, t+H$, are in the state variable. Why? [Hint: look at the transition equations for forecasts and the variables they are forecasting.]

9.3 — What is the difference between the variables x_t, $t = 0, \ldots, T$, and the variables $\tilde{x}_{tt'}$ for $t' = t, \ldots, t+H$?

9.4 — What is a "crossing time"?

9.5 — What is a "Cholesky decomposition" and what is it used for?

9.6 — What state is hidden in the hidden-state Markov model of wind energy? Explain why it is hidden.

9.7 — What class of policy is the parameterized policy in Section 9.4.2? What objective function is used for finding the best set of tuning parameters?

Problem solving questions

9.8 — Try designing a parameterized policy for making decisions under the conditions of the problem in this chapter. You may use anything in the form of rules or parameterized functions. The only limitation is that you are not allowed to optimize over anything (which is to say, you cannot use an $\arg\max_x$ within your policy).

9.9 — Our parameterized lookahead was limited to introducing coefficients in front of forecasts. You may also introduce additive adjustments, such as keeping the energy storage device from getting too close to its capacity (this would allow you to store a burst of wind that exceeds the forecast) or from getting too close to zero (in case you have a dip in the wind). Suggest an alternative parameterization and make an argument why your structure might add value.

10

Supply Chain Management I: The Two-agent Newsvendor Problem

10.1 Narrative

Imagine that a field manager for Amazon has to provide trailers to move freight out of Chicago on a weekly basis. The field manager has access to information that allows him to estimate how many trailers he will need that week, but the actual might be higher or lower. The field manager then makes a request to a central manager for trailers, who then makes a judgment on her own how many trailers to provide, and makes the final decision on the number of trailers that will be provided.

The two managers both work for the same company, but the field manager is far more worried about running out, since he has to do short-term rentals if he runs out of trailers. The central manager, on the other hand, does not want the field manager to run out, but she also does not want him to have excess trailers, since she has to pay for those trailers.

We assume the process unfolds as follows:

Step 1: The field manager observes an initial estimate of how many trailers are needed. This information is private to the field manager.

Step 2: The field manager then requests trailers from the central manager, where he would typically inflate his request to reduce the probability of running out.

Step 3: The central manager then decides how many trailers to give to the field manager, typically reducing the request given the pattern of noticing that the field asked for more than was necessary.

Step 4: The field manager receives the number of trailers granted by the central manager, and then observes the actual need for trailers.

Step 5: The field and central managers compute their performance using their own costs for overage (the unused trailers) and underage (the uncovered demand).

The tension in this problem arises first because the initial estimate of trailers needed is just an estimate, which we may assume is unbiased (that is, it is true on average). The problem is that the field manager has a large cost if he runs out, so his strategy is to overestimate his needs (recall the newsvendor problem in Chapter 3). The central manager, on the other hand, probably has balanced costs for being over and under, and does not want to order too many or too few.

What complicates the problem is the initial estimate given to the field manager. While not perfect, it has valuable information since it will indicate if a day is going to have high or low demand. This means that the central manager has to pay attention to the request made by the field manager, while recognizing that the field manager will make requests that are biased upward. Knowing this, the central manager would tend to use the field manager's request as a starting point, but then reduce this for the final allocation. Not surprisingly, the field manager knows that the central manager will do this, and compensates accordingly.

10.2 Basic model

We will model the problem for both agents, since the information is not the same for both players. Throughout the model, we will be referring to the field manager as q and to the central manager as q'.

10.2.1 State variables

The initial information available to the field manager is the estimate of the number of trailers that will be needed, which we represent using

$$R_{tq}^{est} = \text{the initial estimate of how many trailers are needed.}$$

This initial estimate may be biased, so we introduce an estimate of this bias using

δ_{tq}^{est} = initial estimate of the difference between R_{tq}^{est} and the true demand.

We will also have to estimate how much the central manager reduces the request of the field manager, which we represent using

δ_{tq} = estimate of how much the central manager will reduce the request of the field manager.

Similarly, the central manager will learn the difference between the request made by the field manager, and what the field manager eventually needs, which we represent by

$\delta_{tq'}$ = estimate of the difference between what the field manager requests and what the field eventually needs.

The state variable for each agent is the information they have before they make a decision. For the field manager, the state variable is

$$S_{tq} = (R_{tq}^{est}, \delta_{tq}^{est}, \delta_{tq}).$$

The state variable for the central manager is

$$S_{tq'} = (x_{tqq'}, \delta_{tq'}).$$

where $x_{tqq'}$ is the request made by the field agent q to the central manager q' (introduced next).

10.2.2 Decision variables

The decisions for each agent are given by

$x_{tqq'}$ = the number of trailers that agent q asks for from agent q',

$x_{tq'q}$ = the number of trailers that agent q' gives to agent q, which is what is implemented in the field.

10.2.3 Exogenous information

The exogenous information for the field manager can be thought of as the initial estimate of the trailers needed (although we put that in the state variable):

R^{est}_{tq} = The initial estimate of how many trailers are needed. This estimate is known only to the field agent q.

After making the decision $x_{tqq'}$, we then receive two types of information: what the central manager grants us, and then the actual required demand:

$x_{tq'q}$ = The decision made by the central manager in response to the request of the field manager.

$\hat{R}_{t+1}$ = The actual number of trailers that field manager q ends up needing (this information is available to the central manager as well).

The exogenous information for agent q is then

$$W_{t+1,q} = (x_{tq'q}, \hat{R}_{t+1}).$$

We note in passing that while this information is indexed at time $t+1$, the request granted by the central manager, $x_{tq'q}$, is indexed by t since it depends on information available up through time t. The initial estimate R^{est}_{tq} is new information, but it arrives before the decision is made so it is captured in the state variable for the field agent.

The central manager receives the initial request $x_{tqq'}$ which arrives as exogenous information, but because this is received before she makes her decision, it enters through the state variable for the central manager. The only exogenous information for the central manager is the final demand which might then be used to update beliefs that influence future decisions. This means

$$W_{t+1,q'} = (\hat{R}_{t+1}).$$

10.2.4 Transition function

For the field manager, there are three state variables: R^{est}_{tq}, the bias δ^{est}_{tq} between the estimate R^{est}_{tq} and the actual $\hat{R}_{t+1}$, and the bias δ_{tq}

introduced by the central manager when the field makes a request. The first state variable, R_{tq}^{est}, arrives directly as exogenous information. The biases δ_{tq}^{est} and $\delta_{t,q}$ are updated using

$$
\begin{aligned}
\delta_{t+1,q}^{est} &= (1-\alpha)\delta_{tq}^{est} + \alpha(\hat{R}_{t+1} - R_{tq}^{est}), \\
\delta_{t+1,q} &= (1-\alpha)\delta_{tq} + \alpha(x_{tqq'} - x_{tq'q}),
\end{aligned}
$$

where $0 < \alpha < 1$ is a smoothing factor.

The transition function for the central manager is similar. Again, the decision of the field manager, $x_{tqq'}$ arrives to the state variable exogenously. Then, we update the bias that the central manager estimates in the request of the field manager using

$$
\delta_{t+1,q'} = (1-\alpha)\delta_{t,q'} + \alpha(x_{tqq'} - \hat{R}_{t+1}).
$$

10.2.5 Objective function

We begin by defining:

c_q^o = The unit cost incurred by the field manager for each excess trailer (what the field pays per day for each trailer), also known as the overage cost.

c_q^u = The unit cost incurred by the field manager for each trailer that has to be rented to make up for lack of capacity, also known as the underage cost.

$c_{q'}^o, c_{q'}^u$ = The cost of overage and underage for the central manager.

The costs for each agent are given by

$$
\begin{aligned}
C_{tq}(S_{tq}, x_{tq'q}) &= \text{overage/underage costs for agent } q, \\
&= c_q^o \max\{x_{tq'q} - \hat{R}_{t+1}, 0\} + c_q^u \max\{\hat{R}_{t+1} - x_{tq'q}, 0\}, \\
C_{tq'}(S_{tq'}, x_{tq'q}) &= \text{overage/underage costs for agent } q', \\
&= c_{q'}^o \max\{x_{tq'q} - \hat{R}_{t+1}, 0\} + c_{q'}^u \max\{\hat{R}_{t+1} - x_{tq'q}, 0\}.
\end{aligned}
$$

The performance of both the field and central managers depends on the number of trailers $x_{tq'q}$ that the central manager gives to the field. This decision, however, depends on the decision made by the field manager.

The decisions of the field manager are made with the policy $X_{tq}(S_t|\theta_q)$, where θ_q is one or more tunable parameters that are used to solve

$$\min_{\theta_q} \mathbb{E}\left\{\sum_{t=0}^{T} C_{tq}(S_{tq}, X_{tq}(S_t|\theta_q))|S_0\right\}. \tag{10.1}$$

Similarly, the decisions of the central manager are made with the policy $X_{tq'}(S_t|\theta_{q'})$ where $\theta_{q'}$ is one or more tunable parameters that solve

$$\min_{\theta_{q'}} \mathbb{E}\left\{\sum_{t=0}^{T} C_{tq'}(S_{tq'}, X_{tq'}(S_t|\theta_{q'}))|S_0\right\}. \tag{10.2}$$

The optimization problems in (10.1) and (10.2) have to be solved simultaneously, since both policies have to be simulated at the same time. Of course we could hold $\theta_{q'}$ for the central manager constant while tuning θ_q for the field manager, but ultimately we are looking for a stable local minimum.

10.3 Modeling uncertainty

This problem is data-driven, which means that we react to data as it arrives. There are three types of information, depending on which agent is involved:

- The initial estimate R^{est}_t of the resources required.
- The request $x_{tqq'}$, made by the field manager, that arrives to the central manager. This decision involves logic introduced by the field manager, which may include randomization. This comes as information to the central manager.
- The decision $x_{tq'q}$ made by the central manager that determines the number of trailers given to the field manager. This comes as information to the field manager.
- The final realization $\hat{R}_{t+1}$ of the number of trailers actually required, which is revealed (in this basic model) to both agents.

If we wish to simulate the process, we only need to model the generation of R_t^{est} and $\hat{R}_t$. More precisely, we would have to generate R_t^{est} from one distribution, and the error $\hat{R}_t - R_t^{est}$ from another distribution.

10.4 Designing policies

For our two-agent newsvendor problem, we have to develop policies for each agent. We begin with the policy for the field manager.

10.4.1 Field manager

The field manager starts with an estimate R_t^{est}, but has to account for three factors:

1) The estimate R_t^{est} may have a bias δ^{est} (we cannot be sure about the source of the estimate R_{tq}^{est}). The bias is given by

$$\delta_{tq}^{est} = \mathbb{E}\hat{R}_{t+1} - R_{tq}^{est}.$$

So, if $\delta_{tq}^{est} > 0$ then this means that R_t^{est} is upwardly biased.

2) The true number of trailers needed, $\hat{R}_{t+1}$ is random even once you have factored in the bias. The field manager has a higher cost of having too few trailers than too many, so he will want to introduce an upward bias to reflect the higher cost of being caught short.

3) The central manager has a balanced attitude toward having too many or too few, and knows about the bias of the field manager. As a result, the central manager will typically use the request of the field manager, $x_{tqq'}$, just as the field manager may be adjusting for a possible bias between the estimate R^{est} and the actual $\hat{R}_t$. The field manager knows that the central manager will be making this adjustment, and as a result has to try to estimate it, and counteract it. Since the field manager knows both his request $x_{tqq'}$ and then sees what the central manager provides, the time t observation of the bias is given by

$$\delta_{tq} = x_{tq'q} - x_{tqq'}.$$

We need to use our estimates of the differences between R_t^{est} and $\hat{R}_t$, the difference between $x_{tqq'}$ and $x_{tq'q}$, and the difference between $x_{tqq'}$ and $\hat{R}_t$. We propose a policy for the field manager given by

$$X_{tqq'}(S_t|\theta_q) = R_t^{est} - \delta_{t-1,q}^{est} - \delta_{t-1,q} + \theta_q. \tag{10.3}$$

This policy starts with the initial estimate R_t^{est}, corrects for the bias in this initial estimate using $\delta_{t-1,q}^{est}$, then corrects for the bias from the central manager $\delta_{t-1,q}$, and then finally introduces a shift that can capture the different costs of over and under for the field manager. The parameter θ_q has to be tuned.

Since there is not an imbedded optimization problem (that is, an $\arg\max_x$ or $\arg\min_x$), this is a classic parameterized policy function approximation (PFA).

10.4.2 Central manager

Our policy for the central manager is given by

$$X_{tq'q}(S_t|\theta_{q'}) = x_{tqq'} - \delta_{t-1,q'} + \theta_{q'}. \tag{10.4}$$

Here, we start with the request made by the field manager, subtract our best estimate of the difference between the field manager's request and what was eventually needed, $\delta_{tq'}$, and then add in $\theta_{q'}$ which is a tunable parameter for the central manager, where $\theta_{q'}$ may be negative.

10.4.3 Policy search

We now have two parameterized policies. The tuning of the field policy would be done with the objective function in (10.1), while the tuning of the central policy would be done with the objective function in (10.2). The trick here is that both objectives have to be simulated in parallel, since the policies are interconnected. And while both simulations are running, we are keeping track of the objectives for each agent.

The right way to approach the optimization of the parameters of each agent is to simulate the behavior of both agents simultaneously, but run search algorithms for each agent as if they were separate. The performance of the field agent, for example, would be affected by the behavior of the central agent just as the field agent is affected by the other forms of exogenous information.

This simulation opens the door to some challenging opportunities, such as understanding the effect each agent has on the other. This observation opens up an entire class of extensions where each agent tries to understand how their decision may *change* the behavior of the other agent. We pursue this further in the exercises.

10.5 What did we learn?

- We introduce a basic multiagent problem we call the "two-agent newsvendor problem" where a field agent has to request resources from a central agent. While both agents are supposed to be working together, they each have their own costs of overage (having too many resources) and underage (having too few, producing unsatisfied demands).
- We model information that is private to the field agent $(R^{est}_{tq}, \theta_{tq})$ and the information that is private to the central agent $\theta_{tq'}$.
- The problem introduces the dimension of estimating and anticipating the behavior of the central agent to help the field agent make decisions.
- At each point in time, the field agent has a best estimate of what he wants to order given the estimate R^{est}_{tq} and the history of the central agent adjusting the request. Given the uncertainty and the higher cost of running out than having excess, it is natural to expect that a good policy is to order what we expect to need plus a buffer for uncertainty, so we start by suggesting policies of this form.
- This problem lays the foundation for building in a belief of how the central agent will respond to the adjustment being made by the field agent, since we assume that she ultimately sees the overage or underage.

- While this problem seems quite simple, it lays the foundation for many more complex multiagent resource allocation problems.

10.6 Exercises

Review questions

10.1 — What is known by agent q but unknown by agent q'? Similarly what is known by agent q' but unknown by agent q?

10.2 — There is one piece of information that is made available to both agents. What is this?

10.3 — What is the exogenous information that becomes available to the field agent? What is the exogenous information that becomes available to the central agent?

10.4 — There are three sources of uncertainty in our system. What are they?

Problem solving questions

10.5 — Write out the dynamic models for both the field manager and the central manager. Remember that the decision of one manager becomes exogenous information for the other.

10.6 — What would happen if the field agent simply orders larger and larger quantities? What mechanism could be introduced to the model to minimize this instability?

10.7 — Create an estimate of how the field agent's decision, $x_{tqq'}$ might affect the behavior of the central agent. Then, design a policy that captures this effect so that the decision made by the field agent anticipates the effect of his decision.

10.8 — There is just one piece of information that can be used to create a belief about the policy of another agent. What is that information?

10.9 — Now assume that the field agent gets to sell the resources he obtains from the central agent at a price p_{tq} that changes randomly from one time to the next. This means that the field agent could hold some or all of its resources to a later time period if the price p_{tq} is too low. Expand the model in this chapter to handle this much richer setting. You will need to introduce a new decision variable (how much of the demand to satisfy). Suggest a policy function approximation for making the decision.

Programming questions

These exercises use the Python module *TwoNewsvendor*.[1]

10.10 — Perform a grid search on the bias for the field and central managers. Search over the range $[0, 10]$ (in steps of 1) for the field manager, and $[-11, 0]$ (in steps of 1) for the central manager. Run the game for $N = 30$ time periods, and repeat simulation for 1,000 samples (and take an average). Note that you are adding rewards over the 30 time periods, but averaging over the 1,000 samples. Plot three heat maps for the following:

a) The total reward for the field manager, for each of the combinations of the two biases.

b) The total reward for the central manager, for each of the combinations of the two biases.

c) The total reward for the company (adding the field manager and central manager), for each of the combinations of the two biases. Discuss the differences in the optimal combinations from each of the three perspectives. Each player wants to maximize its reward.

10.11 — Now we are going to use the interval estimation learning policy to learn each of the biases (see Section 4.4). Let θ_q^{IE} be the parameter for the IE policy for the field manager, and let $\theta_{q'}^{IE}$ be the parameter for the IE policy for the central manager. Instead of searching for the best bias, we are going to search for the best parameter to guide the policy for finding the bias.

[1]See http://tinyurl.com/sdagithub

a) Run the Python module varying each learning parameter over the range $(0, 1, 2, 3, 4, 5)$. This means 36 total simulations (over a horizon $N = 20$, and for 1,000 sample paths). Plot the same three heat maps that you did for Exercise 10.10.

b) Compare the behavior of the heat maps from part (a), to the heat maps from Exercise 10.10. Try to explain the behavior of the field and central agents by writing out the policies and thinking about how it should behave.

c) Verify that the direct search for the bias gives the highest overall reward. What are the strengths and weaknesses of each approach in a more realistic setting where the parameters of the problem may change over time?

10.12 — (This exercise requires some modifications to the Python module.) Consider now a two-agent newsvendor problem where the central manager also has some external information about the demand. What he has is a much noisier estimate of the demand (say the noise is, for our spreadsheet data where the demand is always between 20 and 40, three times bigger than the noise from the source communicating with the field manager).
Redefine the bias from the central manager as the quantity that he adds to the estimate he gets. Try a learning approach where the bias he selects is chosen in the interval $[-11; 0]$. Run the program and compare the results with the old learning process. As before, run the game for $N = 30$ time periods, and repeat the simulation for 1,000 samples (and take an average). After $N = 30$ time periods, is the central agent putting more weight on the information coming from the field or from his other external source of information? Why?

10.13 — Consider the case where the field manager is using a learning approach and the central manager is using a punishing strategy. Since he knows the field gets a bigger penalty for providing less than the demand, the central manager will compute the previous field bias (for time $t - 1$) and if it is positive, next round, it will apply a bias twice as big in magnitude and of opposite sign to the field's request. Run this experiment and see what the field's bias will be after the 30 time periods. Comparing this policy with the previous policies, should the central manager employ this strategy?

11

Supply Chain Management II: The Beer Game

11.1 Narrative

This chapter covers a famous game from the 1950's known as the "beer game." This was originally designed by Jay Forrester, a professor at MIT who created the game to illustrate the instabilities of supply chains. The problem involves a linear supply chain where various suppliers move beer from where it is made (the manufacturer) toward the market (the retailer). The beer has to move through several middlemen on its way from point of manufacture to the market.

There are two types of flows:

- The flow of beer - Each case of beer is represented by a penny that moves from manufacturer to retailer.
- The flow of information - Each point in the supply chain replenishes its inventory by making requests for more beer to the next level down.

Each level of the supply chain is known as an *echelon*. Typically there are four to six echelons for each team. The demands at the retailer are fixed in advance, but hidden, in a deck of cards. As the retailer reveals that week's demand, she tries to fill the demand from inventory. The retailer, and every other supplier in the supply chain (other than the manufacturer) then fills out a sheet of paper requesting more inventory.

It is possible that the retailer, or any of the middle suppliers, may not be able to fill the request for more inventory (or the market demand at the retail level). In this case, the unsatisfied demand sits in an order backlog waiting to be filled as new inventory arrives.

After filling orders, everyone (for that supply chain) has to stop and record either their inventory (the number of cases of beer sitting in inventory) or their order backlog. Order backlog carries a penalty of $4 per case. Excess inventory carries a holding cost of $1 per case.

The steps of the process are illustrated in Figure 11.1. There are five steps:

Step 0: Each week, each player will have an inventory (in pennies, each representing a case of beer), and an order, which might be the retail demand (for the retail echelon) or an order placed by the player to their left.

Step 1: Each player tries to take as many cases from their inventory and move it to their left to a point *between* them and the player to the left (do not add the pennies to the inventory of the player to the left). If there is not enough inventory to satisfy the order, cross out the order and replace it with the number of cases that remain to be satisfied (this is the order backlog).

Step 2: Now write out an order of how many cases you want to replenish your inventory, and place it in the area *between* you and the player to your right (the manufacturer places an order on the pile of pennies from which all beer originates).

Step 3: Stop and record on your inventory sheet how much inventory you have. If you have not been able to satisfy an order, you will have no inventory, and you will have orders in your backlog (the unsatisfied orders). If this is the case, record this as your order backlog.

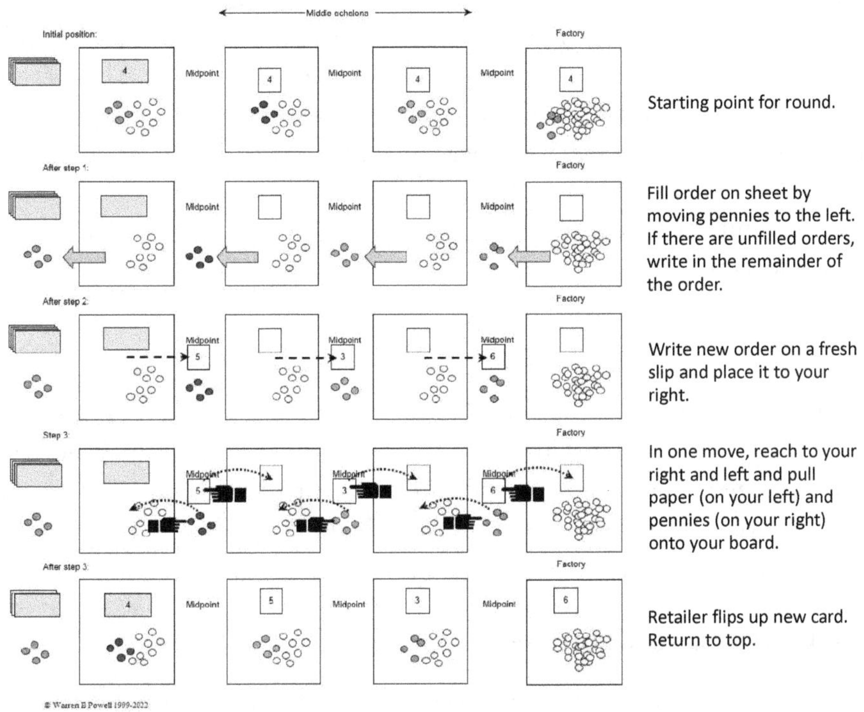

Figure 11.1: Layout of the Princeton version of the beer game.

Step 4: This is the key step: reach out with your left hand and pull the next order slip into your order pile (the sheets of paper), and at the same time reach out with your right hand to pull the pennies coming to you into your inventory. Now you are back to where you were for step 0.

It is very important that everyone make moves at the same time, but they are not allowed to share information (and they should not be looking at the inventories of other players in the same chain). The retailer has to play the role of keeping everyone synchronized.

The complete instructions for a streamlined version of the classic beer game can be downloaded from https://tinyurl.com/PrincetonBeerGame. This version of the game is ideally suited for classes taught at continuous tables to a class with at least 8-10 students (the continuous tables are needed so players can push paper and pennies between each other). Teams should have five or six players (which means five or six intermediate echelons plus the retailer), but no fewer than four. The person running the inventory closest to the factory may run both positions (since the factory does little more than fill orders). The teams do not have to be the same size, and it is fairly easy to extend a chain to add a late-arriving student. It is possible to run the complete game in a 50 minute lecture.

11.2 Basic model

We are going to model a supplier other than one of the endpoints (retailer or beer manufacturer). This model will closely follow the style of the two-agent newsvendor problem in the previous chapter, although there are some adjustments. Before we get started, we have to introduce some new notation for multiagent systems.

11.2.1 Multiagent notation

Before we get started, we need to establish our notational system for who knows what, and the process of sharing information.

We are going to label the different decision-making agents in the supply chain by $\mathcal{Q} = \{1, 2, \ldots, Q\}$. We are going to let $q = 0$ describe the market, which is a source of information but does not make decisions. We are going to let $q = Q$ refer to the manufacturing plant which we assume (at least initially) can always make enough product to fill demand.

We start by defining the state variable for agent q using

S_{tq} = information known to agent q at time t (this may include beliefs).

If agent q acts on agent q', we will use

$x_{tqq'}$ = action by agent q on agent q'.

We note that the decision $x_{tqq'}$ is determined by q, but arrives to q' as information.

An action by q on q' at time t may involve the movement of physical resources, but could also include sending information. The action by q on q' will arrive to q' as an exogenous information process arriving to q' at time $t+1$ (which is where any distortions would be captured), which we write as

$W_{t+1,q,q'}$ = information arriving to agent q' at time $t+1$ from actions taken by agent q (this can be information about resources or one involving the sending or sharing of information from S_{tq}).

Finally, there will be times when agent q needs to create an estimate of something known by agent q'. If we let $S_{tq'}$ represent something known by agent q', we will let

$\overleftarrow{S}_{t,q,q'}$ = the estimate that agent q creates of the information in $S_{tq'}$.

11.2.2 State variables

The state variables for agents $q = 1, \ldots, Q-1$ are:

R^{inv}_{tq} = The inventory left over after iteration t after delivering product to the upstream vendor for agent q.

R^{back}_{tq} = Backordered demand that has not yet been satisfied from inventory.

The manufacturer $q = Q$ is assumed to always have unlimited inventory.

In time we will learn that this is an incomplete statement of the state of the problem, but it is a good starting point.

11.2.3 Decision variables

Agent q has to make two decisions. The first (and most important) is how much to order from the downstream agent $q+1$ which we write as

$x^{req}_{tq,q+1}$ = order placed by supplier q to be passed to supplier $q+1$, made at the order instant in iteration t, which will be received by $q+1$ to be filled in iteration $t+1$.

The second is how much of the request from the upstream agent to fulfill from inventory. We write this as

$x^{fill}_{tq,q-1}$ = how much of the unsatisfied demand R^{back}_{tq} to fill at time t from inventory.

These decisions are constrained for $q = 1, \ldots, Q-1$ by:

$$0 \leq x^{fill}_{tq,q-1} \leq R^{inv}_{tq}, \tag{11.1}$$

$$0 \leq x^{fill}_{tq,q-1} \leq R^{back}_{tq}, \tag{11.2}$$

$$x^{req}_{tq,q+1}, x^{fill}_{tq,q-1} \geq 0. \tag{11.3}$$

Constraint (11.1) reflects the reality that we cannot send inventory to agent $q-1$ that we do not have on hand. Constraint (11.2) says we cannot send inventory to agent $q-1$ that has not been requested. Note that R^{back}_{tq} includes new orders that have not yet been filled.

We then write our decision vector as

$$x_{tq} = (x^{req}_{tq,q+1}, x^{fill}_{tq,q-1}),$$

where our decisions will be made by some policy $X^{\pi}(S_t)$ which we will design later.

In our basic game, we are always going to fill as much of the order from $q-1$ as we can from inventory, so technically $x^{fill}_{tq,q-1}$ is not really a decision since we will just set $x^{fill}_{tq,q-1} = \min\{R^{back}_{tq}, R^{inv}_{tq}\}$. However, it is still an action made by q, and it opens the door for richer behaviors later.

If we are the retail market $q = 0$, then the request $W_{t,0,1} = x^{req}_{t,0,1}$ made to agent $q = 1$ comes from an exogenous information source.

If we are the plant $q = Q$, we always fill the request from $q = Q-1$, so

$$x^{fill}_{t+1,Q,Q-1} = x^{req}_{t,Q-1,Q}.$$

11.2.4 Exogenous information

There are two types of exogenous information for supplier q:

$W^{fill}_{t+1,q+1,q}$ = The amount of product received from supplier $q+1$ in response to the request made at time t but arriving at time $t+1$.

$W^{req}_{t+1,q-1,q}$ = The order made by supplier $q-1$ at time t of supplier q, which would arrive at time $t+1$.

It is important to recognize that the decisions made by agents $q+1$ and $q-1$ come to agent q as exogenous information. This means we could write

$$\begin{aligned} W^{fill}_{t+1,q+1,q} &= x^{fill}_{t,q+1,q}, \\ W^{req}_{t+1,q-1,q} &= x^{req}_{t,q-1,q}. \end{aligned}$$

We can represent the exogenous information for agent q arriving by time $t+1$ using

$$W_{t+1,q} = (W^{fill}_{t+1,q+1,q}, W^{req}_{t+1,q-1,q}).$$

This describes the information process for the middle agents $q = 1, \ldots, Q-1$. The information process $W_{t,0}$ refers to the market where we assume that there is an exogenous source of requests $x^{req}_{t,0,1} = W_{t,0,1}$ that are made to agent 1.

11.2.5 Transition function

Our state variables R^{inv}_{tq} and R^{back}_{tq} for $q = 1, \ldots, Q-1$ evolve according to

$$R^{inv}_{t+1,q} = R^{inv}_{tq} - x^{fill}_{t,q,q-1} + W^{fill}_{t+1,q+1,q}, \tag{11.4}$$

$$R^{back}_{t+1,q} = R^{back}_{tq} - x^{fill}_{t,q,q-1} + W^{req}_{t+1,q-1,q}. \tag{11.5}$$

Equation (11.4) pulls the request $x^{fill}_{t,q,q-1}$ from inventory (it is not allowed to go negative) and then adds incoming inventory $W^{fill}_{t+1,q+1,q}$ from the downstream agent $q+1$ to create the inventory at time $t+1$. Equation (11.5) fills orders that have been requested, held in R^{back}_{tq}, and then adds new orders $W^{req}_{t+1,q-1,q}$ to be filled in period $t+1$.

11.2.6 Objective function

Our objective function for agent q assesses penalties for leftover inventory R_{tq}^{inv} and unsatisfied demand R_{tq}^{back}. Let

$$\begin{aligned} c_q^{inv} &= \text{the unit cost of holding inventory for agent } q, \\ c_q^{back} &= \text{the unit cost of unsatisfied orders for agent } q. \end{aligned}$$

These costs are assessed on inventories and backlogged demands after making decisions to fill a customer order but before new orders have arrived. So, our cost function for agent q is given by

$$C(S_t, x_t) = c^{inv}(R_{tq}^{inv} - x_{t,q,q-1}^{fill}) + c^{back}(R_{tq}^{back} - x_{t,q,q-1}^{fill}).$$

Keep in mind that R_{tq}^{inv} is the current inventory, so $R_{tq}^{inv} - x_{t,q,q-1}^{fill}$ is the remaining inventory after we have filled orders for time t. Similarly, R_{tq}^{back} includes new orders, as well as unfilled orders from previous periods. As a result, $R_{tq}^{back} - x_{t,q,q-1}^{fill}$ is the orders that were not filled right away.

We now search for the best policy using

$$\min_\pi \mathbb{E}\left\{\sum_{t=0}^{T} C_q(S_t, X^\pi(S_t))|S_0\right\}. \tag{11.6}$$

This has to be done for each agent q, assuming that each are self-optimizing. A separate challenge is choosing policies for each agent, which can only use the information available to each agent, but where we still want policies that achieve a global optimality. That is a question beyond the scope of this book.

11.3 Modeling uncertainty

Each agent, other than the agents at the endpoints, has to manage two sources of uncertainty:

- The requests made by the upstream agent, which may be the market, or another agent that is responding in an uncertain way to the demands they face, and the ability of the supply chain to respond to their requests.
- The ability of the upstream agent to fill the orders of the agent.

In other words, the only sources of uncertainty are the market, and the behavior of the agents. The ways in which agents (that are people) interact with each other introduces complex (and uncertain) dynamics. Typically, the game is run where the dynamics of the market is quite modest. Even when it is run this way, human behavior can introduce significant instabilities that have been observed in real supply chains, where they have been given the name "bullwhip effect."

11.4 Designing policies

There are a variety of basic PFA-style policies that we might consider. We start by assuming that we always fill a request up to our available inventory, so

$$x^{fill}_{t,q,q-1} = \min\{x^{req}_{t,q-1,q}, R^{inv}_{tq}\}.$$

We note that as we design different policies, we may have to introduce additional elements to the state variables to meet the information needs of the policy.

11.4.1 Some simple rules

We are going to warm up with some simple ordering rules:

- Request from agent $q+1$ whatever was requested from q in the previous time period:

  $$X^{\pi,req}_{t,q,q+1}(S_{tq}|\theta) = W^{req}_{t-1,q-1,q} + \theta_q.$$

  This policy ignores how much we have in inventory; it is a pure tracking policy. This policy requires that we store the previous request $W^{req}_{t-1,q-1,q}$ in our state variable which becomes $S_{tq} = (R^{inv}_{tq}, R^{back}_{tq}, W_{t-1,q-1,q})$. We then bump it up by θ to protect against uncertainty.

- Request what is needed to meet current and past requests:

  $$X^{\pi,req}_{t,q,q+1}(S_{tq}|\theta) = R^{back}_{tq} + \theta_q.$$

  When we have leftover demands, this policy would be double-counting, which means making multiple requests.

- Target inventory policy:

$$X^{\pi,req}_{t,q,q+1}(S_{tq}|\theta) = \max\{0, \theta^{target}_q - R^{inv}_{tq}\}.$$

This policy seeks to maintain a specified target inventory θ^{target} which is not varied as conditions change.

These are basic parameterized PFAs which are easy to implement, but of course require tuning. At the same time, they are quite simple, and ignore factors such as the history of past orders that have not yet been filled (in fact, each of these policies have fundamental flaws).

Keep in mind that R^{back}_{tq} is the orders by $q-1$ made to q which q has not yet filled. The orders placed by q to $q+1$ that have not yet been filled are given by $R^{back}_{t,q+1}$, but this is not immediately known to agent q. Let:

$\overleftarrow{R}^{back}_{tq,q+1}$ = The estimate of $R^{back}_{t,q+1}$ made by agent q of the back-ordered demands known by $q+1$. These are the unfilled orders that q placed to $q+1$, which is a statistic normally maintained by $q+1$.

Normally information known by one agent (such as $q+1$) cannot be known perfectly by another agent (such as q), but in this case, this is a statistic that q can maintain on its own using

$$\overleftarrow{R}^{back}_{t+1,q,q+1} = \max\{0, \overleftarrow{R}^{back}_{tq,q+1} + x^{req}_{t,q,q+1} - W^{fill}_{t+1,q+1,q}\}.$$

We can use this statistic to suggest an adjusted target inventory policy, where we are adding the unfilled orders captured by $\overleftarrow{R}^{back}_{t+1,q,q+1}$ to our current inventory R^{inv}_{tq}, which we write using:

- Adjusted target inventory policy:

$$x^{req}_{t,q,q+1} = \max\{0, \theta^{target} - (R^{inv}_{tq} + \overleftarrow{R}^{back}_{t+1,q,q+1})\}.$$

This policy is a form of PFA (there is no imbedded optimization), but it reflects supplies that will be arriving in the future.

11.4.2 An anchor-and-adjustment heuristic

In 1989, John Sterman (MIT professor and expert on business dynamics) wrote a paper (Sterman, 1989) applying the principle of "anchor-and-adjustment" developed by Tversky and Kahneman (1974) to the beer game. We are going to outline this idea here.

We begin by defining a set of state variables. The variables we actually use may depend on the policy.

- Physical state variables:

$$\begin{aligned} R_{tq}^{inv} &= \text{Current inventory.} \\ R_{tq}^{back} &= \text{Backlogged demand.} \\ R_{tq}^{transit} &= \text{Current in-transit inventory (we are not capturing how long the inventory has been in transit).} \end{aligned}$$

  The resource state is then $R_{tq} = (R_{tq}^{inv}, R_{tq}^{back}, R_{tq}^{transit})$.

- Information variables:

$$\begin{aligned} F_{t-1,q,q-1} &= \text{Actual fill from } q \text{ to } q-1 \text{ from previous time period,} \\ &= x_{t-1,q,q-1}^{fill}. \\ A_{t-1,q+1,q} &= \text{Actual arrivals to } q \text{ from } q+1 \text{ in previous time period,} \\ &= x_{t-1,q+1,q}^{fill}. \end{aligned}$$

  The information state is then $I_{tq} = (F_{t-1,q-1,q}, A_{t-1,q-1,q})$. With these variables we are "remembering" an activity from the previous time period. Their use depends on the policy.

- Belief state variables:

$$\begin{aligned} \bar{A}_{t,q+1,q} &= \text{Estimated arrival rate of product from agent } q+1. \text{ This is an estimate of the rate at which product is arriving to } q \text{ from } q+1. \\ \overline{F}_{t,q,q-1} &= \text{Estimated fill rate delivered to agent } q-1. \text{ This is an estimate of the rate at which product is being shipped to } q-1. \end{aligned}$$

$\bar{D}_{t,q-1,q}$ = Estimated demand rate from $q-1$. This would equal $\overline{F}_{t,q-1,q}$ if we completely filled every order. This means that $\overline{F}_{t,q-1,q} \leq \bar{D}_{t,q-1,q}$.

The belief state is then $B_{tq} = (\bar{A}_{t,q+1,q}, \overline{F}_{t,q-1,q}, \bar{D}_{t,q-1,q})$. As with I_t, the use of these variables depends on the policy. Later we are going to propose different ways for computing these estimates.

Our complete state variable is then

$$S_{tq} = (R_{tq}, I_{tq}, B_{tq}).$$

The estimated fill rate $\overline{F}_{t,q,q-1}$ may be calculated in any of several ways:

- Reactive

$$\overline{F}_{t,q-1,q} = F_{t-1,q-1,q}.$$

- Stable

$$\overline{F}_{t,q-1,q} = \theta_q^{trgt-fill}$$

where $\theta_q^{trgt-fill}$ is a target fill rate set by agent q.

- Regressive expectations

$$\overline{F}_{t,q-1,q} = (1-\gamma)\overline{F}_{t-1,q-1,q} + \gamma\theta_q^{trgt-fill}$$

for a specified smoothing factor $0 \leq \gamma \leq 1$.

- Adaptive expectations

$$\overline{F}_{t,q-1,q} = (1-\gamma)\overline{F}_{t-1,q-1,q} + \gamma\overline{F}_{t,q-1,q}.$$

The principle of "anchor-and-adjustment" applied to this setting is to choose an "anchor" which specifies how much we expect we should be ordering on average, with an "adjustment" to reflect current conditions.

- Basic refill policy - We can use any of the methods for computing $\overline{F}$ to obtain the policy

$$X^{\pi,req}_{t,q,q+1}(S_{tq}) = \overline{F}_{t,q-1,q}.$$

- Anchor and adjust policy - We are going to use our estimated order rate $\overline{F}_{t,q-1,q}$ as our "anchor," which is what we expect we should order, but we are going to make adjustments based on our inventory on hand, and inventory in-transit. We represent these adjustments using:

$$\begin{aligned}\delta R^{inv}_{tq} &= \text{Adjustment based on the current inventory } R^{inv}_{tq}.\\ \delta R^{transit}_{tq} &= \text{Adjustment based on the current in-transit inventory } R^{transit}_{tq}.\end{aligned}$$

  We can use these to create an "anchor and adjustment" policy given by

$$X^{\pi,req}_{t,q,q+1}(S_{tq}|\theta_q) = \max\{0, \overline{F}_{t,q-1,q} + \delta R^{inv}_{tq} + \delta R^{transit}_{tq}\}.$$

  So now we have to design adjustment mechanisms. A possible strategy for δR^{inv}_t might be

$$\delta R^{inv}_{tq} = \theta^{inv}_q (R^{inv-trgt}_q - R^{inv}_{tq}),$$

  where θ^{inv}_q is a smoothing factor and the target inventory $R^{inv-trgt}$ are tunable parameters.

  A possible strategy for $\delta R^{transit}_{tq}$ might be

$$\delta R^{transit}_{tq} = \theta^{transit}(R^{transit-trgt}_q - R^{transit}_{tq}).$$

  Our vector of tunable parameters would then be

$$\theta_q = (\theta^{inv}_q, R^{inv-trgt}_q, \theta^{transit}_q, R^{transit-trgt}_q).$$

  These parameters have to be tuned for each agent q.

The anchor and adjustment policy was motivated by human behavior, rather than any justification that it would be near optimal. An advantage is that it is simple, transparent and intuitive. The challenge is always the tunable parameters, and in particular the targets $R^{inv-trgt}$ and $R^{transit-trgt}$, since these are presented as static parameters, when in fact they really need to respond to conditions. Finding optimal values of these parameters does not mean we have an optimal policy; just one that is optimal within this class.

11.4.3 A lookahead policy

We first posed a stochastic lookahead policy in Equation (7.4), but we replicate it here for ease of reference:

$$X^{DLA}(S_t) = \arg\min_{x_t \in \mathcal{X}} \Bigg(C(S_t, x_t) +$$

$$\tilde{E}_{\tilde{W}_{t,t+1}} \left\{ \min_{\tilde{\pi}} \mathbb{E}_{\tilde{W}_{t,t+2},\ldots,\tilde{W}_{tT}} \left\{ \sum_{t'=t+1}^{T} C(\tilde{S}_{tt'}, \tilde{X}_t^{\tilde{\pi}}(\tilde{S}_{tt'})) | \tilde{S}_{t,t+1} \right\} | S_t, x_t \right\} \Bigg). \tag{11.7}$$

Equation (11.7) can be particularly daunting. Figure 11.2 illustrates each of the elements in the policy using a basic decision tree (all of this is for a single agent q which we suppress). There is a set of decisions x_t that emanate from the first decision node S_t, after which we take an expectation over the random information in $\tilde{W}_{t,t+1}$. After that, we use an approximate "lookahead policy" $\tilde{X}^{\tilde{\pi}}(\tilde{S}_{tt'})$ for each decision node $\tilde{S}_{tt'}$ in our lookahead model, where we typically simplify the state variable in some way. We also approximate information arriving in the future using $\tilde{W}_{tt'}$, either by using a deterministic lookahead model, or a simulated set of possible outcomes.

This equation can be thought of as consisting of two elements:

- We first enumerate each possible decision x_{tq}.
- Then, we simulate the effects of this decision using a sample of any random information, while making decisions using an approximate "lookahead policy" that is designated $\tilde{X}^{\tilde{\pi}}(\tilde{S}_{tt'})$.

To use this in our supply chain setting, we have to simulate the behavior of the other agents, realizing that a) we do not know the initial conditions $R_{tq'}$ for $q' \neq q$, and b) we do not know how the other agents are making decisions.

To handle our lack of knowledge of the starting conditions, we have to view these as random variables and sample from a distribution (this is buried in the first $\tilde{E}_{\tilde{W}_{t,t+1}}$). We note that it is possible for an agent to have, for example, no inventory and a substantial backorder. We

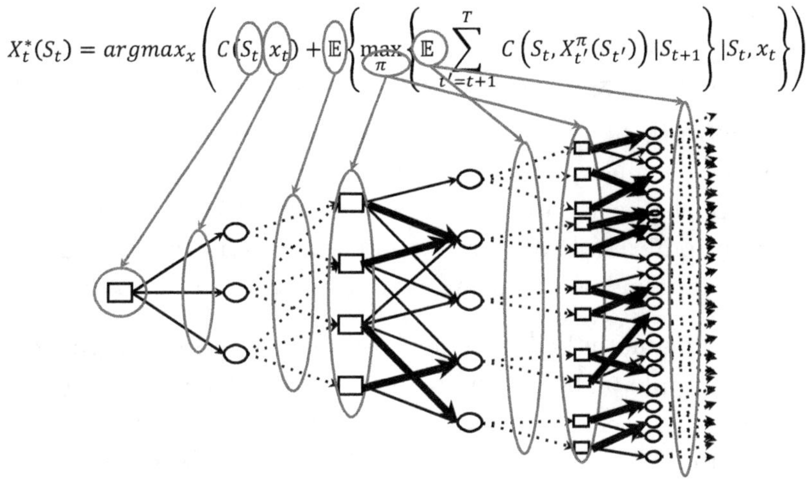

Figure 11.2: Illustration of lookahead policy as a decision tree (from Powell, 2022).

might be able to guess that this is the case if we see that the time to fill the orders we are placing are taking a long time to be filled.

We then have to simulate the unknown policies. While we are trying to build a very sophisticated stochastic lookahead policy for agent q at time t, we suggest using the much simpler policies we have suggested earlier, not just for the other agents, but also for agent q in future time periods.

So, given these approximations, would a stochastic lookahead policy outperform one of the simpler policies we sketched above? This would be a nice research question, but the lookahead policy overcomes a major limitation of the simpler parametric policies. Specifically, the lookahead policy naturally captures the complex state of this system such as the history of previous orders as well as any forecasts of future events. The parametric policies are suited to stationary problems, while the lookahead naturally adapts to behavior that may be highly nonstationary.

11.5 Extensions

There are many ways we can modify this problem. Some ideas include:

1) We have to handle situations where the orders from upstream agents are much bigger (or perhaps smaller) than what we have seen in the past, hinting at a systematic change in demand. We can introduce estimates of the potential growth in future demands to handle unexpected changes in the upstream demand.

2) We can maintain beliefs about how upstream agents might behave. For example, it helps agent q if agent $q + 1$ maintains generous inventories. We can encourage inventory building by introducing noise in our own requests which then increase the estimate that agent $q + 1$ has of the uncertainty in the orders placed by agent q.

3) Each agent reacts to outages, and responds by maintaining higher inventories. An agent q could introduce some noise in their orders to $q + 1$ so that $q + 1$ will maintain higher inventories so that the orders from q are more likely to be filled.

4) Much of the sensitivity of the game is due to the high penalty for stocking out versus holding inventory (recall that it costs \$4 per case per day of backlogged orders, and \$1 per case per day to hold inventory). Try changing the stockout cost from \$4 to \$1, and then to \$0.50.

11.6 What did we learn?

- We describe a simple multiagent problem called the "beer game" that was invented in the 1950s. To model the problem, we introduce additional notation to capture the knowledge of each agent, and the transfer of information between agents. This can be thought of as a series of two-agent newsvendor problems, with the twist that excess inventory is held to the next time period, as are unsatisfied demands.

- Readers are pointed to a streamlined version of the classic beer game developed by the author at Princeton University.

- We introduce notation that captures what each agent knows, including an estimate by one agent of information known to another agent.
- We model a decision by one agent coming as exogenous information to another agent.
- We begin with some simple PFA policies where agents adapt to basic information on how much they need to order.
- We then summarize a famous "anchor and adjustment" policy suggested by two psychologists, which is another form of PFA.
- Finally we sketch a direct lookahead policy that depends on an agent q simulating the behavior of other agents.

11.7 Exercises

Review questions

11.1 — What is the state of an intermediate agent?

11.2 — What decisions can be made by each agent?

11.3 — What are the sources of exogenous information for each intermediate agent?

11.4 — What sources of uncertainty affect the behavior of the game?

11.5 — Explain in words what is meant by an "anchor and adjustment" policy? Was it the expectation of its designers that this might be a good policy?

Problem solving questions

11.6 — Offer a critique of the simple rules suggested in Section 11.4.1.

11.7 — Imagine that there are occasional, but infrequent, shifts in the demand from the market to much higher or lower levels. Design a policy that understands that these shifts may happen, which means that the rest of the supply chain also has to adapt. How would your policy respect to the invariable periods of shortages of product?

11.8 — Section 11.4.3 provides a rough sketch of a lookahead policy. Fill in the details by writing out an implementation in detail.

12

Ad-click Optimization

12.1 Narrative

Companies advertising on internet sites such as Google have to bid to get their ads in a visible position (that is, at the top of the list of sponsored ads). When a customer enters a search term, Google identifies all the bidders who have listed the same (or similar) search terms in their list of ad-words. Google then takes all the matches, and sorts them in terms of how much each participant has bid, and runs an auction. The higher the bid, the more likely your ad will be placed near the top of the list of sponsored ads, which increases the probability of a click. Figure 12.1 is an example of what is produced after entering the search terms "hotels in baltimore md."

Companies such as Google and Facebook have raised the process of identifying the ads that are likely to attract the most clicks to a high science, where auctions allow them to extract the maximum value from an ad. At the same time, advertisers are able to learn whether a user clicks on the ad, and even whether the user then makes a purchase. Needless to say, identifying the best ad is very important.

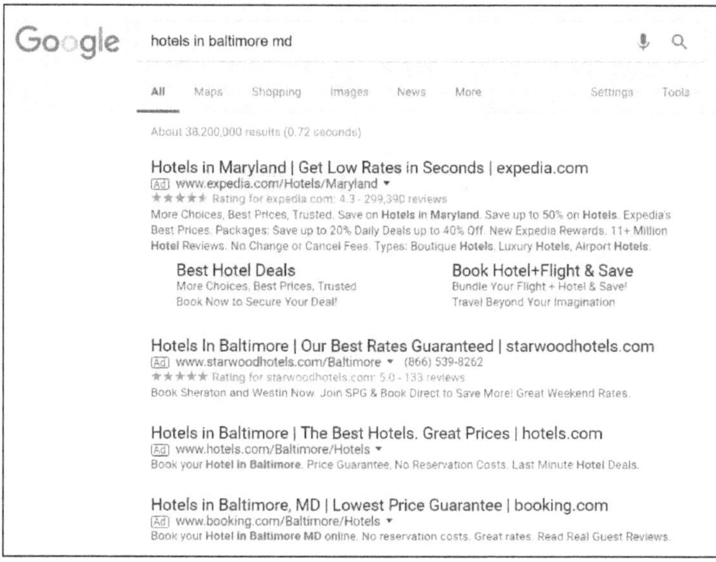

Figure 12.1: Sample of displayed ads in response to an ad-word search.

If a customer clicks on the ad, there is an expected return that reflects the average amount a customer spends when they visit the website of the company. The problem is that we do not know the bid response curve. Figure 12.2 reflects a family of possible response curves. Our challenge is to try out different bids to learn which curve is correct.

We begin by assuming that we can adjust the bid after every auction, which means we only learn a single response (the customer did or did not click on the link). It is possible that the customer looked at a displayed link and decided not to click it, or our bid may have been so low that we were not even in the list of displayed ads.

Our challenge is to design a policy for setting the bids. The goal is to maximize the net revenue, including what we make from selling our products or services, minus what we spend on ad-clicks.

12.2 Basic model

We are going to assume that we use some sort of parameterized model to capture the probability that a customer clicks on an ad. At a minimum this probability will depend on how much we bid for an ad - the more

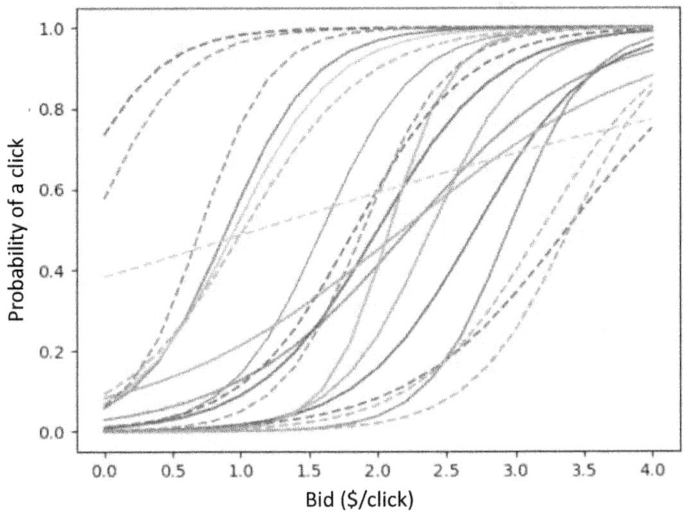

Figure 12.2: Possible instances of the probability of an ad getting a click given the bid.

we bid, the higher the ad will appear in the sponsored ad list, which increases the likelihood that a customer will click on it. Let $K^n = 1$ if the nth customer clicks on the ad. Let

$$P^{click}(\theta_k, x) \quad = \quad Prob[K^{n+1} = 1|\theta = \theta_k, x]$$

where $Prob[K^{n+1} = 1|\theta = \theta_k, x]$ will be described by a logistic function given by

$$Prob[K^{n+1} = 1|\theta = \theta_k, x^n, H^n] \quad = \quad \frac{e^{\theta_k^{const,n} + \theta_k^{bid,n} x^n}}{1 + e^{\theta_k^{const,n} + \theta_k^{bid,n} x^n}}. \quad (12.1)$$

This function is parameterized by $\theta = (\theta^{const}, \theta^{bid})$. We do not know what θ is, but we are going to assume that it is one of a sampled set $\Theta = \{\theta_1, \ldots, \theta_K\}$.

12.2.1 State variables

The initial state S^0 includes:

$$\begin{aligned}\Theta &= \{\theta_1, \ldots, \theta_K\} \\ &= \text{the set of possible values that } \theta \text{ may take.} \\ \bar{R}^0 &= \text{Initial estimate of revenue earned when a customer clicks on a link.}\end{aligned}$$

The dynamic state variables S^n includes:

$$\begin{aligned}p_k^n &= \text{Probability that the true } \theta = \theta_k. \\ p^n &= (p_k^n)_{k=1}^K. \\ \bar{R}^n &= \text{Estimate of revenue earned from an ad-click after } n \text{ auctions.}\end{aligned}$$

Our dynamic state variable, then, is

$$S^n = (\bar{R}^n, p^n).$$

Note that we can create a point estimate of θ after n observations using

$$\bar{\theta}^n = \sum_{k=1}^K p_k^n \theta_k,$$

but this is a statistic that we can compute from the information in S^n, so we do not put $\bar{\theta}^n$ in the state variable.

12.2.2 Decision variables

Our only decision variable is the bid which we define as

$$x^n = \text{the bid (in \$ per click) for the } (n+1)^{st} \text{ auction.}$$

As before, we let $X^\pi(S^n)$ be our generic policy that gives us the bid x^n as a function of the information available to us, represented by S^n, which means that we would write

$$x^n = X^\pi(S^n).$$

We assume that the policy enforces any constraints, such as making sure that the bid is not negative or something too large.

12.2.3 Exogenous information

In our initial model, we only observe the results of a single auction, which we model using:

$$K^{n+1} = \begin{cases} 1 & \text{If the customer clicks on our ad,} \\ 0 & \text{otherwise.} \end{cases}$$
$$\hat{R}^{n+1} = \text{Revenue earned from the } n+1^{st} \text{ auction.}$$

This means our complete exogenous information variable is

$$W^{n+1} = (\hat{R}^{n+1}, K^{n+1}).$$

12.2.4 Transition function

The transition function for this problem will seem a lot more complicated than others in this volume, which is because we are updating beliefs about the uncertainty in the parameter vector θ. We need to emphasize that all the transition equations can be coded relatively easily.

We are going to update our estimated revenue when a customer clicks on the ad using:

$$\bar{R}^{n+1} = \begin{cases} (1-\alpha^{lrn})\bar{R}^n + \alpha^{lrn}\hat{R}^{n+1} & \text{If } K^{n+1} = 1, \\ \bar{R}^n & \text{otherwise.} \end{cases} \tag{12.2}$$

Thus, we only update our estimated revenue when we get a click. The parameter α^{lrn} is a smoothing parameter (sometimes called a "learning rate") between 0 and 1 that we fix in advance.

We next address the updating of the probabilities p_k^n. We let H^n be the history of states, decisions and exogenous information

$$H^n = (S^0, x^0, W^1, S^1, x^1, \ldots, W^n, S^n, x^n).$$

We use this to write

$$p_k^n = Prob[\theta = \theta_k | H^n].$$

The way to read the conditioning on the history H^n is "p_k^n is the probability $\theta = \theta_k$ given what we know after n observations." We then use Bayes theorem to write

$$
\begin{aligned}
p_k^{n+1} &= Prob[\theta = \theta_k | W^{n+1}, H^n] \\
&= \frac{Prob[K^{n+1}|\theta = \theta_k, H^n] Prob[\theta = \theta_k | H^n]}{Prob[K^{n+1}|H^n]}.
\end{aligned} \tag{12.3}
$$

Remember that the history H^n includes the decision x^n which, given a policy for making these decisions, is directly a function of the state S^n (which in turn is a function of the history H^n). We now use our logistic curve in Equation (12.1) to write

$$
\begin{aligned}
Prob[K^{n+1} = 1|\theta = \theta_k, H^n] &= Prob[K^{n+1} = 1|\theta = \theta_k, x^n] \\
&= \frac{e^{\theta_k^{const} + \theta_k^{bid} x^n}}{1 + e^{\theta_k^{const} + \theta_k^{bid} x^n}}.
\end{aligned} \tag{12.4}
$$

We then note that

$$
Prob[\theta = \theta_k | H^n] = p_k^n. \tag{12.5}
$$

Finally, we note that the denominator can be computed using

$$
Prob[K^{n+1}|H^n] = \sum_{k=1}^{K} Prob[K^{n+1}|\theta = \theta_k, H^n] p_k^n. \tag{12.6}
$$

Our use of a sampled representation of the possible outcomes of θ is saving us here. Even if θ has only two dimensions (as it is here, but only for now), performing a two-dimensional integral over a multivariate distribution for θ would be problematic.

Equations (12.4) - (12.6) allow us to calculate our Bayesian updating equation for the probabilities in (12.3). Equations (12.2)-(12.3) make up our transition function

$$
S^{n+1} = S^M(S^n, x^n, W^{n+1}).
$$

12.2.5 Objective function

We begin by writing the single period profit function as

$$
C(S^n, x^n, W^{n+1}) = (\hat{R}^{n+1} - x^n) K^{n+1},
$$

which means we make nothing if the customer does not click on the ad ($K^{n+1} = 0$). If the customer does click on the ad ($K^{n+1} = 1$), we receive revenue given by $\hat{R}^{n+1}$, but we also have to pay what we bid for the ad-click, given by our bid x^n.

We will end up taking the expected contribution, which we write as

$$\mathbb{E}\{C(S^n, x^n, W^{n+1})|S^n\} \quad = \quad \mathbb{E}\{(\hat{R}^{n+1} - x^n)K^{n+1}|S^n\}. \quad (12.7)$$

There are three random variables hidden in the expectation: θ, with distribution p^n (contained in S^n), K^{n+1}, where

$$P^{click}(\theta, x) = Prob[K^{n+1} = 1|\theta, x],$$

and $\hat{R}^{n+1}$, which we observe from some unknown distribution if $K^{n+1} = 1$, and where $\hat{R}^{n+1} = 0$ if $K^{n+1} = 0$ (we do not get any revenue if the customer does not click on the ad).

We can then break the expectation into three nested expectations:

$$\mathbb{E}\{(\hat{R}^{n+1} - x^n)K^{n+1}|S^n\} \quad = \quad \mathbb{E}_\theta \mathbb{E}_{K|\theta} \mathbb{E}_{\hat{R}}\{(\hat{R}^{n+1} - x^n)K^{n+1}|S^n\}.$$

We start by taking the expectation over $\hat{R}$ where we just use $\mathbb{E}\{\hat{R}^{n+1}|S^n\} = \bar{R}^n$ (remember that $\bar{R}^n$ is in the state variable S^n), which allows us to write

$$\mathbb{E}\{(\hat{R}^{n+1} - x^n)K^{n+1}|S^n\} \quad = \quad \mathbb{E}_\theta \mathbb{E}_{K|\theta}\{(\bar{R}^n - x^n)K^{n+1}|S^n\}.$$

Next we are going to take the expectation over K^{n+1} for a given θ using

$$\mathbb{E}_{K|\theta}\{(\bar{R}^n - x^n)K^{n+1}|S^n\} \quad = \quad (\bar{R}^n - x^n)P^{click}(\theta, x).$$

where we have used the fact that $(\bar{R}^n - x^n)K^{n+1} = 0$ if $K^{n+1} = 0$.

Finally we take the expectation over θ using

$$\mathbb{E}_\theta\{(\bar{R}^n - x^n)P^{click}(\theta, x^n)|S^n\} \quad = \quad \sum_{k=1}^{K}(\bar{R}^n - x^n)P^{click}(\theta = \theta_k, x^n)p_k^n.$$

We are going to let $\bar{C}(S^n, x)$ be the expected contribution, which is to say

$$\bar{C}(S^n, x) \quad = \quad \mathbb{E}_\theta \mathbb{E}_{K|\theta}\{(\bar{R}^n - x^n)K^{n+1}|S^n\}.$$

Our objective function can now be written as

$$\max_\pi \mathbb{E}_{S^0} \mathbb{E}_{W^1,\ldots,W^n|S^0} \left\{ \sum_{n=0}^{N} C(S^n, X^\pi(S^n), W^{n+1})|S_0 \right\}.$$

Note that the conditioning on S_0 is how we communicate our prior $p_k^0 = Prob[\theta = \theta_k]$ to the model. As before, we would approximate the expectation by averaging over simulated samples of the true value of θ, and the observed clicks K^n and revenues R^n.

12.3 Modeling uncertainty

We have three forms of uncertainty: the ad-click K^{n+1}, the revenue we receive $\hat{R}^{n+1}$ if $K^{n+1} = 1$, and then the true value of θ. We are going to assume that we simply observe $\hat{R}^{n+1}$ from a real datastream, which means we do not need a formal probability model for these random variables. We assume that K^{n+1} is described by our logistic function

$$\begin{aligned} P^{click}(\theta, x) &= P[K^{n+1} = 1|\theta, x = x^n] \\ &= \frac{e^{\theta^{const} + \theta^{bid} x}}{1 + e^{\theta^{const} + \theta^{bid} x}}, \end{aligned} \tag{12.8}$$

but it is important to recognize that this is just a fitted curve. The values of K^{n+1} are observed from data, which means we have no guarantee that the distribution precisely matches our logistic regression.

Finally, we assume that $\theta \in \Theta = \{\theta_1, \ldots, \theta_K\}$ which is also an approximation. There are ways to relax the requirement of a sampled set, but the logic becomes somewhat more complicated without adding much educational value.

12.4 Designing policies

We are going to explore three policies for learning:

- Pure exploitation - Here we always place the bid that appears to be the best given our current estimates.
- An excitation policy - We introduce exploration into our exploitation policy by adding in a random noise term which forces the system to explore in regions near the areas we think are best (this is popular in engineering where states and decisions are continuous).

- A value of information policy - We are going to maximize the value of information from placing a bid and learning the outcome.

12.4.1 Pure exploitation

The starting point of any online policy should be pure exploitation, which means doing the best that we can. To compute this we start by using

$$\begin{aligned} \mathbb{E}\{\hat{R}^{n+1}K^{n+1}\} &= \mathbb{E}\{\hat{R}^{n+1}|K^{n+1}=1\}Prob[K^{n+1}=1|\theta=\theta_k] \\ &= \bar{R}^n P^{click}(\theta,x). \end{aligned}$$

To find the best bid, we find (after a bit of algebra) the derivative with respect to the bid x

$$\frac{d\bar{C}(x)}{dx} = (\bar{R}^n - x)\frac{dP^{click}(\theta,x)}{dx} - P^{click}(\theta,x)$$

where

$$\frac{dP^{click}(\theta,x)}{dx} = \frac{\theta_1 e^{-\theta_0-\theta_1 x}}{(1+e^{-\theta_0-\theta_1 x})^2}.$$

Now we want to find the bid x^* where

$$\left.\frac{d\bar{C}(x)}{dx}\right|_{x=x^*} = 0.$$

Figure 12.3 shows $\frac{d\bar{C}(x|\theta)}{dx}$ versus the bid x, showing the behavior that it starts positive and transitions to negative. The point where it is equal to zero would be the optimal bid, a point which can be found numerically quite easily. Let

$$X^{explt}(S^n) = \text{the bid } x^* \text{ satisfying } d\bar{C}(x)/dx = 0.$$

This means that we have to run a numerical algorithm to compute the policy. This is a greedy policy which falls in the CFA class, but without any tunable parameters.

12.4.2 An excitation policy

A potential limitation of our pure exploitation policy is that it ignores the value of trying a wider range of bids to help with the process of

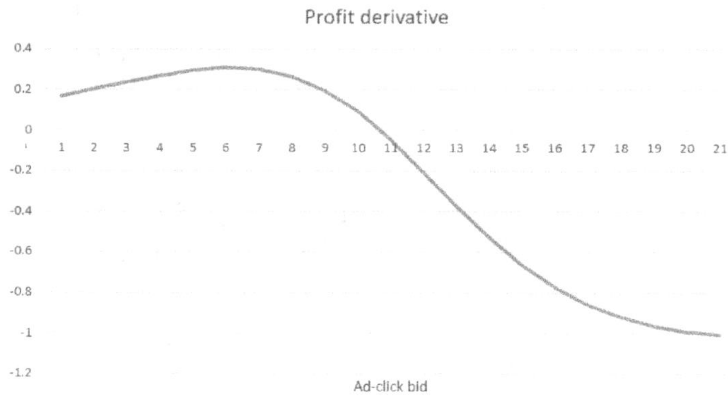

Figure 12.3: Derivative of ad-click profit function versus the bid.

learning the correct values of θ. A popular strategy is to add a noise term, known in engineering as "excitation," giving us the policy

$$X^{excite}(S^n|\rho) \quad = \quad X^{explt}(S^n) + \varepsilon(\rho)$$

where $\varepsilon(\rho) \sim N(0, \rho^2)$. In this policy, ρ is our tunable parameter which controls the amount of exploration in the policy. If it is too small, then there may not be enough exploration. If it is too large, then we will choose bids that are far from optimal, possibly without any benefit from learning.

12.4.3 A value of information policy

The pure exploitation and excitation policies we just introduced are both relatively simple. Now we are going to consider a policy that maximizes the value of information in the future. This seems like a reasonable idea, but it requires that we think about how information now affects what decision we *might* make in the future, and this will be a bit more difficult.

Our exploitation policy assumes that the estimated parameters θ^n after n experiments is the correct value, and chooses a bid based on this estimate. Now imagine that we bid $x^n = x$ and observe K^{n+1} and $\hat{R}^{n+1}$, and use this information to get an updated estimate of θ^{n+1} as well as $\bar{R}^{n+1}$. We can then use these updated estimates to make a better decision. We want to choose the bid x that gives us the greatest improvement in the value of information from a decision, recognizing that we do not know the outcome of $W^{n+1} = (\hat{R}^{n+1}, K^{n+1})$ until we actually place the bid.

Let $\theta^{n+1}(x^n, W^{n+1})$ be the updated estimate of θ assuming we bid $x^n = x$ and observe $W^{n+1} = (\hat{R}^{n+1}, K^{n+1})$. This is a random variable, because we are thinking about placing a bid $x^n = x$ for the $n+1^{st}$ auction, but we have not yet placed the bid, which means we have not yet observed W^{n+1}.

To simplify our analysis, we are going to assume that the random variable $K^{n+1} = 1$ with probability $P^{click}(\theta, x)$ and $K^{n+1} = 0$ with probability $1 - P^{click}(\theta, x)$. We are then going to assume that our estimate of the revenue we receive from an ad-click has stabilized, which means that $\bar{R}^{n+1} \approx \bar{R}^n$.

We can think of this as an approximate lookahead model, where $\bar{R}^n$ does not change. We would then write our exogenous information in our lookahead model as

$$\tilde{W}^{n,n+1} = \tilde{K}^{n,n+1},$$

where the double-superscript $(n, n+1)$ means that this is the information in a lookahead model created at time n, looking at what might happen at time $n+1$. The random variable $\tilde{K}^{n,n+1}$ is the ad-click that we are simulating *might* happen in our lookahead model, rather than the actual observation of whether someone clicked on the ad. Just remember that we use tilde for any variable in our lookahead model, and these variables will be indexed by n (the time at which we are initiating the lookahead model), and $n+1$ (since we are looking one time period forward in the lookahead model).

We next use our updating Equation (12.3) for the probabilities $p_k^n = Prob[\theta = \theta_k | H^n]$. We can write these updated probabilities as $\tilde{p}_k^{n,n+1}(\tilde{K}^{n,n+1})$ to capture the dependence of the updating on $\tilde{K}^{n,n+1}$ (Equation (12.3) is written for $\tilde{K}^{n,n+1} = 1$). Since $\tilde{K}^{n,n+1}$ can take on two outcomes (0 or 1) we will have two possible values for $\tilde{p}_k^{n,n+1}(\tilde{K}^{n,n+1})$.

Now imagine that we perform our pure exploitation policy $X^{explt}(S^n|\theta^n)$ that we described above, but we are going to do it in our approximate lookahead model (this is where we ignore changes in $\bar{R}^n$). Let $\tilde{S}^{n,n+1}$ represent our state in the lookahead model given by

$$\tilde{S}^{n,n+1}(\tilde{K}^{n,n+1}) = (\bar{R}^n, \tilde{p}^{n,n+1}(\tilde{K}^{n,n+1})).$$

Remember - since $\tilde{K}^{n,n+1}$ is a random variable (we are still at time n), $\tilde{S}^{n,n+1}(\tilde{K}^{n,n+1})$ is also a random variable, which is why we write its explicit dependence on the outcome $\tilde{K}^{n,n+1}$.

The way to think about this lookahead model is as if you are playing a game (such as chess) where you think about a move (for us, that would be the bid x^n) and then, before you make the move, think about what might happen in the future. In this problem, our future only has two outcomes (whether or not a customer clicks on the ad), which means two possible values of $\tilde{S}^{n,n+1}$, which produce two sets of updated probabilities $\tilde{p}^{n,n+1}(\tilde{K}^{n,n+1})$.

Finally, this means that there will be two values of the optimal myopic bid (using our pure exploitation policy) $X^{explt}(\tilde{S}^{n,n+1})$. The expected contribution we would make in the future is then given by $\tilde{C}(\tilde{S}^{n,n+1}, \tilde{x}^{n,n+1})$ where $\tilde{x}^{n,n+1}$ (this is the decision we are thinking of making in the future) is given by

$$\tilde{x}^{n,n+1} = X^{explt}(\tilde{S}^{n,n+1}).$$

This means there are two possible optimal decisions, which means two different values of the expected contribution $\tilde{C}(\tilde{S}^{n,n+1}, X^{explt}(\tilde{S}^{n,n+1}))$. For compactness, let's call these $\tilde{C}^{n,n+1}(1)$ (if $\tilde{K}^{n,n+1} = 1$) and $\tilde{C}^{n,n+1}(0)$ (if $\tilde{K}^{n,n+1} = 0$). Think of these as the expected contributions that *might* happen in the future given what we know now. Finally we can take the expectation over $\tilde{K}^{n,n+1}$ to obtain the expected contribution of placing a bid $x^n = x$ right now, which we can compute using

$$\bar{C}^n(x) = \sum_{k=1}^{K} \left(P^{click}(\theta = \theta_k, x)\tilde{C}^{n,n+1}(1) + (1 - P^{click}(\theta = \theta_k, x))\tilde{C}^{n,n+1}(0)\right)p_k^n.$$

Our policy, then, is to pick the bid x that maximizes $\bar{C}^n(x)$. Assume that we discretize our bids into a set $\mathcal{X} = \{x_1, \ldots, x_M\}$. Our value of information policy would be written as

$$X^{VoI}(S^n) = \arg\max_{x \in \mathcal{X}} \bar{C}^n(x).$$

We note that this is a direct lookahead approximation (DLA) class of policy.

Value of information policies are quite powerful. They are harder to compute, but do not have any tunable parameters. Imagine, for example, doing this computation when there is more than two outcomes. For example, if we had not made our simplification of holding $\bar{R}^n$ constant, we would have to recognize that this state variable is also changing.

We note only in passing that we have run many comparisons of different learning policies, and the one-step lookahead value of information often works quite well. We used this setting because it made the derivations much simpler.

A word of caution is in order. Learning problems where the outcome is 0 or 1 are problems where a single experiment provides very little information. Instead, it is better to assume that we make our decision (that is, set the bid) and then observe it for, say, M auctions. This means that $\tilde{K}^{n,n+1}$ might now be a number between 0 and M. The number M becomes a tunable parameter, and the calculations just became a little more complex (we have to sum over $M+1$ realizations rather than just two), but this approach can work quite well.

12.5 Extension: Customers with simple attributes

Assume that we know the location of a customer down to a region or the nearest major city, which we designate by L. If we think that the behavior of each region is different, we could index θ by θ_ℓ if the customer is from location $L = \ell$. This means that if there are 1,000 locations, then we have to estimate 1,000 models, which means 1,000 values of $\theta = (\theta^{const}, \theta^{bid})$.

An alternative approach would be to specify a model of the form

$$Prob^n[K^{n+1} = 1|\theta] \quad = \quad \frac{e^{U(x,L|\theta)}}{1 + e^{U(x,L|\theta)}}. \tag{12.9}$$

where we are now going to use as our utility function

$$U(x, L|\theta) = \theta^{const} + \theta^{bid} x + \sum_{\ell=1}^{L} \theta_\ell^{loc} I_{\ell=L}.$$

This is a more compact model because we now assume that the constant term θ^{const} and bid coefficient θ^{bid} do not depend on the location. Instead, we are just adding a shift θ_ℓ^{loc}. So, we still have 1,000 parameters to estimate (the location coefficients), but before we had 2,000 parameters to estimate - θ_ℓ^{const} and θ_ℓ^{bid} for each location $\ell \in \{1, \ldots, L\}$.

12.6 What did we learn?

- This is another pure learning problem (our diabetes problem in Chapter 4 was a pure learning problem) but this time we are using a nonlinear belief model, with a sampled model for the unknown parameter θ that determines the response to price.
- The transition function includes the Bayesian updating of the beliefs about the probabilities p_k^n that the unknown parameter θ is equal to a specific value θ_k.
- There are three forms of uncertainty: whether someone will click on an ad given the bid price; the revenue earned from clicking on the ad (for example, did the customer purchase the product), and the uncertainty about the market response captured by the unknown parameter θ.
- We illustrate a pure exploitation policy, an excitation policy (which simply randomizes the recommended price from the exploitation policy), and a knowledge gradient policy that maximizes the value of information.

12.7 Exercises

Review questions

12.1 — What probabilistic model is assumed for the random variable K^n giving whether a customer clicked on the ad or not?

12.2 — What probabilistic model did we assume for the unknown (and therefore uncertain) parameter vector θ?

12.3 — What probability distribution did we assume for the revenue $\hat{R}^{n+1}$ we receive when the customer clicks on an ad?

12.4 — Give the equation numbers of the equations that make up the transition function.

12.5 — What probabilistic information is in the initial state S^0?

12.6 — What is accomplished by adding the noise term $\varepsilon(\rho)$ to create the excitation policy? What specific parameter(s) does this help us identify?

12.7 — Describe in words the logic behind the value of information policy. What is the value if learning whether or not a customer clicks on the ad does not change what we are going to bid?

Problem solving questions

12.8 — Recommender system part I - Belief model - You are going to help design a recommender system that recommends products to advertise when a customer is scrolling through a website. Since the customer has to sign in, we can identify the nth customer by a vector of attributes $a = a^n$ that includes:

$$
\begin{aligned}
a_1 &= \text{Gender (2 types).} \\
a_2 &= \text{Age range } (0-10, 11-20, \ldots, 70-100) \text{ (8 types).} \\
a_3 &= \text{Device type (smartphone, laptop, tablet) (3 types).} \\
a_4 &= \text{Region (200).} \\
a_5 &= \text{Unique ID (email address) (100 million).}
\end{aligned}
$$

Imagine that we are recommending text articles. Assume that the article we recommend for the nth customer has attributes $b = b^n$ that includes:

b_1 = News, sports, arts, business, cooking, real-estate (6 types).

b_2 = Subcategory: if news, then international, national (by country), regional (region within a country); if sports, then by sport, and then by team (or athlete); and so on (a total of 500).

b_3 = Source (website, newspaper, ...) (5 sources).

b_4 = Author (2,000).

b_5 = Unique ID for article (6 million).

We would like to estimate:

$$P(b^n|a^n) \;=\; \text{Probability that the } n^{th} \text{ customer with attribute } a^n \text{ clicks on the link of an article with attribute } b^n.$$

When customer a^n arrives, we are going to assume that we have to choose a news article from a set $\mathcal{B}^n$, which is the set of articles available when the nth customer arrives (this set changes over time). We would like to choose an article with attribute $b \in \mathcal{B}^n$ that maximizes the probability that our customer will click on this news article. Our policy has to choose a particular article with attribute b^n.

Ideally, we want $P(b_5^n|a_5^n)$ which is the probability that user a_5^n would select article b_5^n, but there are too many users and too many articles to get reasonable estimates of this probability. If we only consider the elements a_1, a_2, a_3 and a_4, there would be 9,600 combinations, with an average of approximately 10,000 people for each of these first four elements. Below, we are going to assume we just use a_1 and a_2, which means 16 types of people.

We are going to create a set of features $\mathcal{F}$ which are constructed from the elements of a and b that we wish to consider. We are just going to use the elements $\{a_1, a_2, b_1, b_2, b_3\}$ from which we are going to construct a set of feature variables $\phi_f(a, b),\ f \in \mathcal{F}$. Since these five elements are all categorical, the most elementary features are indicator variables. For example, for the gender attribute a_1 we have two genders from which we create two features:

$$\phi_{male}(a) = \begin{cases} 1 & \text{If } a_1 = male, \\ 0 & \text{Otherwise.} \end{cases}$$

$$\phi_{female}(a) = \begin{cases} 1 & \text{If } a_1 = female, \\ 0 & \text{Otherwise.} \end{cases}$$

If we restrict ourselves to these elementary features, we would have one feature for each possible value for each element of the attributes a_1, a_2, b_1, b_2, b_3.

Our process begins when the first customer logs in with attribute vector a^1, at which point we have to decide the attributes of an article b^1 to display to this user, and then observe Y^1, where $Y^1 = 1$ if the customer clicks on the article or 0 otherwise. This information is used to create an updated state S^1, after which we observe customer a^2.

If a^n is the attributes of the nth customer, then our decision is to choose b^n using what we know, which we designate by S^n. Our goal is to model this problem and design a policy $B^\pi(S^n)$ that determines b^n.
Our first challenge is to develop a belief model:

a) If we use a lookup table belief model for $P(b|a)$ using the attributes $\{a_1, a_2, b_1, b_2, b_3\}$, how many parameters are we trying to estimate?

b) Instead, consider using a logistic regression. First define a utility function

$$U(a, b|\theta) = \sum_{f \in \mathcal{F}} \theta_f \phi_f(b|a), \tag{12.10}$$

where $\mathcal{F}$ is the set of elementary features that we can construct from the elements $\{a_1, a_2, b_1, b_2, b_3\}$. Now create a logistic regression model for the probability of clicking on an article using

$$P(Y = 1|a, b, \theta) = \frac{e^{U(a,b|\theta)}}{1 + e^{U(a,b|\theta)}}. \tag{12.11}$$

What is the dimensionality of the vector θ assuming that we just use elementary indicator variables?

c) Recognizing that the number of parameters in the parametric model in part (b) is much smaller than the number of parameters in the lookup table model in part (a), why would anyone use a lookup table belief model instead of a parametric model such as the logistic regression? Discuss the pros and cons of each type of belief model.

d) We now need to estimate θ. Assume we generate a sample of possible values of the vector θ which we represent as $\{\theta_1, \ldots, \theta_k, \ldots, \theta_K\}$, where each θ_k is a vector with element θ_{kf}, $f \in \mathcal{F}$. Start with the prior probability $p_k^0 = 1/K$. Next assume that we observe the attributes of the first customer a^1, and then we make the decision to display an article with attribute b^1 (this is our decision variable). Assuming you know p_k^n, write out Bayes theorem to compute p_k^{n+1} after observing a customer with attribute a^{n+1}, and then choosing an article with attribute b^{n+1} after which you observe the outcome $Y^{n+1} = 1$.

12.9 — Recommender system part II - System model - Now we are going to model all five elements of the problem.

a) Give the elements of the pre-decision state S^n and the post-decision state $S^{b,n}$.

b) There are two forms of exogenous information in this process. What are they?

c) Write out the sequence of states (pre- and post-), decisions and the different forms of exogenous information starting with what you know at time 0 and proceeding up to (but excluding) the arrival of the third customer. Write them in the order that they occur, with proper indexing (e.g. n versus $n+1$).

d) Write out the equations representing the transition function.

e) Write out the objective function to find the best policy $B^\pi(S^n)$ (without specifying the type of policy).

12.10 — Recommender system part III - Policy design - Finally we are going to try to design policies. Assume that we have $K = 20$ possible values of θ.

a) Begin by assuming that we know that $\theta = \theta_k$. Write out a pure exploitation policy where we choose the attribute $b \in \mathcal{B}^n$ that maximizes the probability of being chosen, given that $\theta = \theta_k$.

b) Next assume that we do not know that $\theta = \theta_k$. Instead, $\theta = \theta_k$ with probability p_k^n. Rewrite your policy from part (a) where you have to treat θ as a random variable. You will need to insert an expectation somewhere.

c) The policy in (b) might be viewed as being too expensive to compute. You can simplify it by replacing the random variable θ with its expectation

$$\bar{\theta}^n = \mathbb{E}^n \theta_k = \sum_{k=1}^{K} \theta_k p_k^n.$$

Rewrite your policy from part (b) using this point estimate. Assume that you are only looking at articles where the probability of clicking on an article is greater than 0.5. How do you think the probability of clicking on an article computed using the point estimate in (c) would compare to the estimate provided using the expectation in (b)?

d) The interval estimation policy uses, say, the 95th percentile of the estimate of the value of a choice. Let ρ be the desired percentile, and assume that it has to be rounded to 0.05 (because we have chosen $K = 20$ possible values for θ). Show how to design a policy that chooses the attribute vector b that maximizes the ρth probability (rather than the point estimate), and give the objective function for finding the best value of ρ to maximize the total number of ad-clicks.

13

Blood Management Problem

13.1 Narrative

The problem of managing blood inventories serves as a particularly elegant illustration of a resource allocation problem. We are going to start by assuming that we are managing inventories at a single hospital, where each week we have to decide which of our blood inventories should be used for the demands that need to be served in the upcoming week.

We have to start with a bit of background about blood. For the purposes of managing blood inventories, we care primarily about blood type and age. Although there is a vast range of differences in the blood of two individuals, for most purposes doctors focus on the eight major blood types: $A+$ ("A positive"), $A-$ ("A negative"), $B+$, $B-$, $AB+$, $AB-$, $O+$, and $O-$. While the ability to substitute different blood types can depend on the nature of the operation, for most purposes blood can be substituted according to Table 13.1.

A second important characteristic of blood is its age. The storage of blood is limited to six weeks, after which it has to be discarded. Hospitals need to anticipate if they think they can use blood before it hits this limit, as it can be transferred to blood centers which monitor inventories at different hospitals within a region. It helps if a hospital

Table 13.1: Allowable blood substitutions for most operations, 'X' means a substitution is allowed.

	Recipient							
Donor	$AB+$	$AB-$	$A+$	$A-$	$B+$	$B-$	$O+$	$O-$
$AB+$	X							
$AB-$	X	X						
$A+$	X		X					
$A-$	X	X	X	X				
$B+$	X				X			
$B-$	X	X			X	X		
$O+$	X		X		X		X	
$O-$	X	X	X	X	X	X	X	X

can identify blood it will not need as soon as possible so that the blood can be transferred to locations that are running short.

One mechanism for extending the shelf-life of blood is to freeze it. Frozen blood can be stored up to 10 years, but it takes at least an hour to thaw, limiting its use in emergency situations or operations where the amount of blood needed is highly uncertain. In addition, once frozen blood is thawed it must be used within 24 hours.

13.2 Basic model

13.2.1 State variables

We can model the blood problem as a heterogeneous resource allocation problem. We are going to start with a fairly basic model which can be easily extended with almost no notational changes. We begin by describing the attributes of a unit of stored blood using

$$b = \begin{pmatrix} b_1 \\ b_2 \end{pmatrix} = \begin{pmatrix} \text{blood type } (A+, A-, \ldots) \\ \text{age (in weeks)} \end{pmatrix},$$

$$\mathcal{B} = \text{set of all blood attribute types.}$$

We will limit the age to the range $0 \leq b_2 \leq 6$. Blood with $b_2 = 6$ (which means blood that is already six weeks old) is no longer usable. We assume that decision epochs are made in one-week increments. Blood inventories are represented using

$$\begin{aligned} R_{tb} &= \text{units of blood of type } b \text{ available to be assigned or held at time } t, \\ R_t &= (R_{tb})_{b \in \mathcal{B}}. \end{aligned}$$

The attributes of demand for blood are given by

$$\begin{aligned} a &= \begin{pmatrix} a_1 \\ a_2 \\ a_3 \end{pmatrix} = \begin{pmatrix} \text{blood type of patient} \\ \text{surgery type: urgent or elective} \\ \text{is substitution allowed?} \end{pmatrix}, \\ \mathcal{A} &= \text{set of all attribute types for blood demands.} \end{aligned}$$

The attribute a_3 captures the fact that there are some operations where a doctor will not allow any substitution. One example is childbirth, since infants may not be able to handle a different blood type, even if it is an allowable substitute. For our basic model, we do not allow unserved demand in one week to be held to a later week.

We then define the demand for blood using

$$\begin{aligned} D_{ta} &= \text{the number of units of blood required for patients with attribute } a \text{ at time } t, \\ D_t &= (D_{ta})_{a \in \mathcal{A}}. \end{aligned}$$

The state variables are given by

$$S_t = (R_t, D_t).$$

13.2.2 Decision variables

We act on blood resources with decisions given by

$$\begin{aligned} d &= \text{a type of decision, which includes decisions to give blood to a patient with attribute } a \in \mathcal{A}, \text{ or to do nothing and hold the blood, which we represent by } d^{\phi}, \\ \mathcal{D} &= \text{the set of all possible decisions,} \\ &= \mathcal{A} \cup d^{\phi}. \end{aligned}$$

We then let

$$\begin{aligned} x_{tbd} &= \text{number of units of blood with attribute } b \text{ that we act} \\ &\quad \text{on with a decision of type } d \text{ at time } t, \\ x_t &= (x_{tbd})_{b\in\mathcal{B},d\in\mathcal{D}}. \end{aligned}$$

The feasible region $\mathcal{X}_t$ is defined by the following constraints:

$$\sum_{d\in\mathcal{D}} x_{tbd} = R_{tb}, \quad b \in \mathcal{B}, \tag{13.1}$$

$$\sum_{b\in\mathcal{B}} x_{tbd} \leq \hat{D}_{td}, \quad d \in \mathcal{D}, \tag{13.2}$$

$$x_{tbd} \geq 0. \tag{13.3}$$

13.2.3 Exogenous information

The information that arrives after we have made a decision is given by the blood donations which we represent using

$$\begin{aligned} \hat{R}_{t+1,b} &= \text{number of new units of blood of type } b \text{ donated between } t \text{ and } t+1, \\ \hat{R}_{t+1} &= (\hat{R}_{t+1,b})_{b\in\mathcal{B}}. \end{aligned}$$

The new demands for blood are modeled using

$$\begin{aligned} \hat{D}_{t+1,a} &= \text{units of demand with attribute } a \text{ that arose between } t \text{ and } t+1, \\ \hat{D}_{t+1} &= (\hat{D}_{t+1,a})_{a\in\mathcal{A}}. \end{aligned}$$

Our exogenous information variable would be

$$W_{t+1} = (\hat{R}_{t+1}, \hat{D}_{t+1}).$$

13.2.4 Transition function

Blood that is held simply ages one week, but we limit the age to six weeks. Blood that is assigned to satisfy a demand can be modeled as being moved to a blood-type sink, denoted, perhaps, using $b_{t,1} = \phi$ (the null blood type). The blood attribute transition function $r^M(b_t, d_t)$ is given by

$$b_{t+1} = \begin{pmatrix} b_{t+1,1} \\ b_{t+1,2} \end{pmatrix} = \begin{cases} \begin{pmatrix} b_{t,1} \\ \min\{6, b_{t,2}+1\} \end{pmatrix}, & d_t = d^\phi, \\ \begin{pmatrix} \phi \\ - \end{pmatrix}, & d_t \in \mathcal{D}. \end{cases}$$

To represent the transition function, it is useful to define

$$\delta_{b'}(b,d) = \begin{cases} 1 & b_t^x = b' = b^M(b_t, d_t), \\ 0 & \text{otherwise}, \end{cases}$$
$$\Delta = \text{matrix with } \delta_{b'}(b,d) \text{ in row } b' \text{ and column } (b,d).$$

We note that the attribute transition function is deterministic. A random element would arise, for example, if inspections of the blood resulted in blood that was less than six weeks old being judged to have expired. The resource transition function can now be written as

$$R^x_{tb'} = \sum_{b\in\mathcal{B}} \sum_{d\in\mathcal{D}} \delta_{b'}(b,d) x_{tbd},$$
$$R_{t+1,b'} = R^x_{tb'} + \hat{R}_{t+1,b'}.$$

In matrix form, these would be written as

$$R^x_t = \Delta x_t, \tag{13.4}$$
$$R_{t+1} = R^x_t + \hat{R}_{t+1}. \tag{13.5}$$

The demands D_{t+1} are simply observed from the new demands $\hat{D}_{t+1}$, so we write this as

$$D_{t+1} = \hat{D}_{t+1}.$$

Figure 13.1 illustrates the transitions that are occurring in week t. We either have to decide which type of blood to use to satisfy a demand (Figure 13.1a), or to hold the blood until the following week. If we use blood to satisfy a demand, it is assumed lost from the system. If we hold the blood until the following week, it is transformed into blood

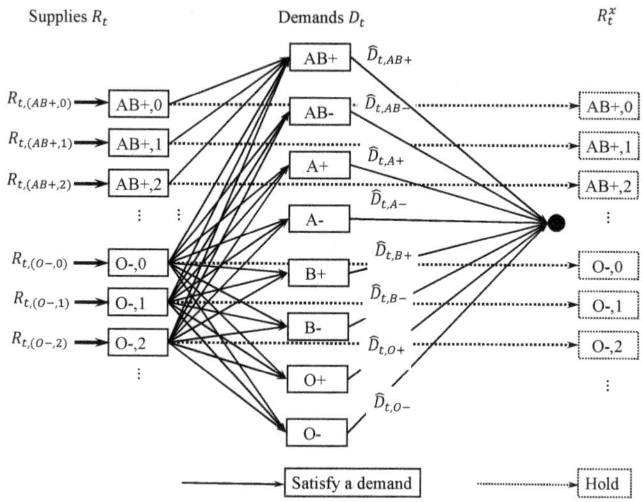

13.1a - Assigning blood supplies to demands in week t. Solid lines represent assigning blood to a demand, dotted lines represent holding blood.

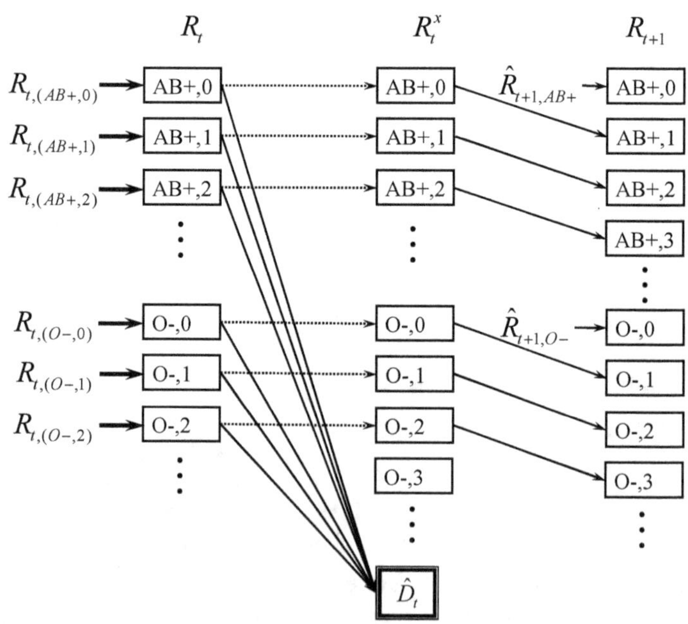

13.1b - Holding blood supplies until week $t + 1$.

Figure 13.1: The assignment of different blood types (and ages) to known demands in week t (13.1a), and holding blood until the following week (13.1b).

that is one week older. Blood that is six weeks old may not be used to satisfy any demands, so we can view the bucket of blood that is six weeks old as a sink for unusable blood (the value of this blood would be zero). Note that blood donations are assumed to arrive with an age of 0.

13.2.5 Objective function

There is no real "cost" to assigning blood of one type to demand of another type (we are not considering steps such as spending money to encourage additional donations, or transporting inventories from one hospital to another). Instead, we use the contribution function to capture the preferences of the doctor. We would like to capture the natural preference that it is generally better not to substitute, and that satisfying an urgent demand is more important than an elective demand.

For example, we might use the contributions described in Table 13.2. Thus, if we use $O-$ blood to satisfy the needs for an elective patient with $A+$ blood, we would pick up a -\$10 contribution (penalty since it is negative) for substituting blood, a +\$5 for using $O-$ blood (something the hospitals like to encourage), and a +\$20 contribution for serving an elective demand, for a total contribution of +\$15.

Table 13.2: Contributions for different types of blood and decisions

Condition	Description	Value
if $d = d^{\phi}$	Holding	0
if $b_1 = b_1$ when $d \in \mathcal{D}$	No substitution	0
if $b_1 \neq b_1$ when $d \in \mathcal{D}$	Substitution	-10
if $b_1 =$ O- when $d \in \mathcal{D}$	O- substitution	5
if $d_2 =$ Urgent	Filling urgent demand	40
if $d_2 =$ Elective	Filling elective demand	20

The total contribution (at time t) is finally given by

$$C_t(S_t, x_t) = \sum_{b \in \mathcal{B}} \sum_{d \in \mathcal{D}} c_{tbd} x_{tbd}.$$

As before, let $X_t^\pi(S_t)$ be a policy (some sort of decision rule) that determines $x_t \in \mathcal{X}_t$ given S_t. We wish to find the best policy by solving

$$\max_{\pi \in \Pi} \mathbb{E} \sum_{t=0}^{T} C_t(S_t, x_t). \tag{13.6}$$

13.3 Modeling uncertainty

The sources of uncertainty in this problem are blood donations, and the arrival of new surgeries requiring blood donations. Some issues we need to consider when modeling this uncertainty include:

- Not only is there randomness in the number of units of blood being donated, but also the type of blood.
- There are both day-of-week and seasonal patterns to blood donations, as well as responses to appeals which, of course, represent a decision.
- The arrival of new surgeries can come in bursts due to weather or violence.
- There is a consistent mismatch between the types of people who donate blood, and the types of people who need surgeries, which emerges in differences in the distribution of blood types.
- A major issue is managing the substitution of blood types. A lot of attention is given to the ability to use $O-$ blood for anyone, but there are different types of substitution for all blood types.

13.4 Designing policies

We are going to start with a basic myopic policy, and then transition to one that depends on approximating the value of blood inventories in the future.

13.4.1 A myopic policy

The most obvious way to solve this problem is a simple myopic policy, where we maximize the contribution at each point in time without regard to the effect of our decisions on the future. We can obtain a family of myopic policies by adjusting the one-period contributions.

For example, our bonus of \$5 for using $O-$ blood (in Table 13.2), is actually a type of myopic policy. We encourage using $O-$ blood since it is generally more available than other blood types. By changing this bonus, we obtain different types of myopic policies that we can represent by the set Π^M, where for $\pi \in \Pi^M$ our decision function would be given by

$$X_t^\pi(S_t) = \arg\max_{x_t \in \mathcal{X}_t} \sum_{b \in \mathcal{B}} \sum_{d \in \mathcal{D}} c_{tbd} x_{tbd}. \tag{13.7}$$

The optimization problem in (13.7) is a simple linear program. Searching over policies in the optimization problem given by Equation (13.6) means searching over different values of the bonus for using $O-$ blood.

13.4.2 A VFA policy

As a traditional dynamic program, the optimization problem posed in Equation (13.6) is quite daunting. The state variable S_t has $|\mathcal{A}| + |\mathcal{B}| = 8 \times 6 + 8 \times 2 \times 2 = 80$ dimensions. The random variables $\hat{R}$ and $\hat{D}$ also have a combined 80 dimensions. The decision vector x_t has $27 + 8 = 35$ dimensions.

It is natural to use value function approximations to determine the allocation vector x_t using

$$x_t^n = \arg\max_{x_t \in \mathcal{X}_t^n} \left(C_t(S_t^n, x_t) + \overline{V}_t^{x,n-1}(R_t^x)\right), \tag{13.8}$$

where $R_t^x = R^M(R_t, x_t)$ is given by Equation (13.4) and where $\mathcal{X}_t^n$ is defined by the constraints (13.1)-(13.3). The key constraint is (13.1) which limits the availability of blood supplies of each type.

The first (and most important) challenge we face is identifying an appropriate approximation strategy for $\overline{V}_t^{x,n-1}(R_t^x)$. A simple and effective approximation is to use separable, piecewise linear approximations, which is to say

$$\overline{V}_t^x(R_t^x) = \sum_{b \in \mathcal{B}} \overline{V}_{tb}^x(R_{tb}^x),$$

where $\overline{V}_{tb}^x(R_{tb}^x)$ is a scalar, piecewise, linear function in the post-decision inventory R_{tb}^x for each blood type b.

It is easy to show that the value function is concave (as well as piecewise linear), so each $\overline{V}_{tb}^x(R_{tb}^x)$ should also be concave. Without loss of generality, we can assume that $\overline{V}_{tb}^x(R_{tb}^x) = 0$ for $R_{tb}^x = 0$, which means the function is completely characterized by its set of slopes. We can write the function using

$$\overline{V}_{tb}^{n-1}(R_{tb}^x) = \left(\sum_{r=1}^{\lfloor R_{tb}^x \rfloor} \bar{v}_{tb}^{n-1}(r-1) + (R_{tb}^x - \lfloor R_{tb}^x \rfloor)\bar{v}_{tb}^{n-1}(\lfloor R_{tb}^x \rfloor) \right), \quad (13.9)$$

where $\lfloor R \rfloor$ is the largest integer less than or equal to R. As we can see, this function is determined by the set of slopes $(\bar{v}_{tb}^{n-1}(r))$ for $r = 0, 1, \ldots, R^{max}$, where R^{max} is an upper bound on the number of resources of a particular type.

The way we estimate the slopes in $\overline{V}_t(R_t)$ is to create the objective function for the problem at time t

$$\tilde{V}_t(S_t) = \max_{x_t \in \mathcal{X}_t^n} \left(C_t(S_t^n, x_t) + \overline{V}_t^{x,n-1}(R_t^x) \right). \quad (13.10)$$

When we solve this linear program, we obtain estimates of the marginal value of an additional unit of blood of type a given by R_{ta}^n. Call this value $\hat{v}_{ta}^n$, which is immediately available from any linear programming package (and we get this for every type of blood a all at the same time).

Alternatively, we could calculate the marginal value more accurately by creating a perturbed resource vector $R_{ta}^{n+} = R_{ta}^n + 1$. Let $\mathcal{X}_t^{n+}(a)$ be the feasible region (made up of Equations (13.1)-(13.3)) where we use R_{ta}^{n+} instead of R_{ta}^n for a single attribute a, and let $\tilde{V}_{ta}^+(S_t)$ be the same as $\tilde{V}_t(S_t)$ except with the feasible region $\mathcal{X}_{ta}^{n+}$ with the perturbed resource R_{ta}^{n+} instead of R_{ta}^{b}. We can then find the marginal values $\hat{v}_{ta}^{n}$ using

$$\hat{v}_{ta}^{n} = \tilde{V}_{ta}^{+}(S_t) - \tilde{V}_t(S_t).$$

Note that we have to calculate $\tilde{V}_{ta}^{+}(S_t)$ for each a (while with dual variables, we get the entire set of marginal values all at once).

We then use $\hat{v}_{ta}^{n}$ to update the *previous post-decision value function approximation* $\bar{v}_{t-1,a}^{x,n}$ which is done with

$$\bar{v}_{t-1,a}^{x,n}(R_{t-1,a}^{x,n}) = (1-\alpha)\bar{v}_{t-1,a}^{x,n-1}(R_{ta}^{x,n}) + \alpha\hat{v}_{ta}^{n}.$$

We can show that the slopes $\bar{v}_{ta}^{x,n}(R_{ta}^{x,n})$ decrease as $R_{ta}^{x,n}$ increases, so it helps when we maintain this. We can do this with methods such as the CAVE or Leveling algorithms (see Powell, 2022, Section 18.3).

Assuming we can estimate this function, the optimization problem that we have to solve (Equation (13.8)) is the fairly modest linear program shown in Figure 13.2. As with Figure 13.1, we have to consider both the assignment of different types of blood to different types of demand, and the decision to hold blood.

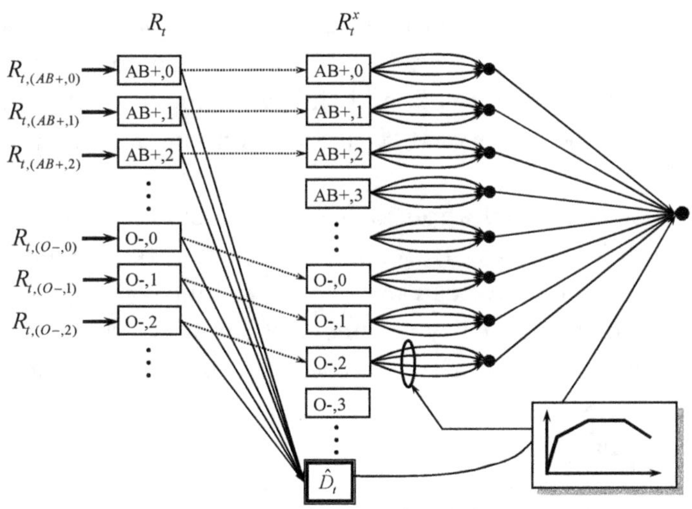

Figure 13.2: Network model for time t with separable, piecewise linear value function approximations.

To simplify the figure, we have collapsed the network of different demand types into a single aggregate box with demand $\hat{D}_t$. This network would actually look just like the network in Figure 13.1a. The decision to hold blood has to consider the value of a type of blood (including its age) in the future, which we are approximating using separable, piecewise linear value functions.

Here, we use a standard modeling trick that converts the separable, piecewise linear value function approximations into a series of parallel links from each node representing an element of R_t^x into a supersink. Piecewise linear functions are not only easy to solve (we just need access to a linear programming solver), they are easy to estimate. In addition, for many problem classes (but not all) they have been found to produce very fast convergence with high quality solutions.

With this decision function, we are going to use a method called *approximate value iteration* where we simulate forward iteratively through time periods $t = 0, \ldots, T$. Let $n = 1, \ldots, N$ be the iteration counter, where we follow a sample path of the exogenous information W_t^n, $t = 0, \ldots, T$ (these might be pulled from history, or sampled from a distribution). At time t, iteration n, we use Equation (13.8) to make a decision x_t^n when we are in state S_t^n. We then observe W_{t+1}^n and use our transition function (Equations (13.4)-(13.5)) for the transition for R_t^n to R_{t+1}^n. When in state $S_t^n = (R_t^n, \hat{D}_t^n)$, we use our VFA policy in Equation (13.8) to compute x_t^n, and then we compute $\hat{v}_t^n$ to update the slopes $\bar{v}_{t-1}^{x,n}$. We then observe W_{t+1}^n (which contains $\hat{D}_{t+1}^n$) to transition to state S_{t+1}^n.

For most operational applications, this problem would be solved over a finite horizon (say, 10 weeks) giving us a recommendation of what to do right now. We can use the value function approximations $\overline{V}_t^x(R_t^x)$ to simulate the policy a number of times, which can be used to produce a form of probabilistic forecast of future inventories.

13.5 Extensions

This is a rich and complex resource allocation problem which can be extended in a number of ways. Below are a few examples.

1) We assume that any demands that are not satisfied at time t are lost. Imagine that we have emergency surgeries that have to be satisfied, and elective surgeries that can be delayed to a later time period. Write out the state variable for the new problem.

2) Assume that elective surgeries can be delayed. Consider using a value function approximation that is piecewise linear and separable in the blood inventories (this is the VFA suggested above), along with piecewise linear and separable VFAs for the amount of held demand (by blood type). We use the dual variables for the blood inventory to update the VFA for blood supplies. How might we update the VFA for the held demand?

3) Include the presence of blood that has been frozen, and the decision to freeze blood, where frozen blood that is not used has to be discarded. This means that we have to recognize that the amount of blood needed for a surgery is unknown before the surgery, when the decision to thaw the blood has to be made.

4) A hospital might require weekly deliveries of blood from a community blood bank to make up for systematic shortages. Imagine that a fixed quantity (e.g., 100 units) of blood arrives each week, but the amount of blood of each type and age (the blood may have already been held in inventory for several weeks) might be random.

5) We presented a model that focused only on blood inventories at a single hospital. We can handle multiple hospitals and distribution centers by simply adding a location attribute, and providing for a decision to move blood (at a cost) from one location to another.

This model can also be applied to any multiproduct inventory problem where there are different types of product and different types of demands, as long as we have the ability to choose which type of product is assigned to each type of demand. We also assume that products are not reusable; once the product is assigned to a demand, it is lost from the system.

13.6 What did we learn?

- We introduce a multidimensional resource allocation problem that has an extremely large state space, but offers the structure of concavity that we can exploit in the approximation of value functions.
- We show how to model a multiattribute resource allocation problem, which makes it quite easy to introduce additional attributes.
- We illustrate a basic myopic policy where the downstream impact of decisions made now is ignored.
- We then describe a VFA policy where we exploit the natural concavity of the problem to suggest an approximation based on separable, piecewise linear value function approximations.
- Our VFA policy would have to be implemented on a rolling basis, so our policy is actually a stochastic DLA, using a VFA policy for the lookahead policy. This parallels our use of dynamic programming to solve the deterministic shortest path problem which represented an approximation of our stochastic, dynamic shortest path problem in Chapter 6.

13.7 Exercises

Review questions

13.1 — What is the dimensionality of the state variable S_t?

13.2 — What is the dimensionality of the decision vector x_t?

13.3 — What are the source(s) of uncertainty?

13.4 — Describe the nature of the costs in the objective function? Where do these come from?

13.5 — What is the limitation of a purely myopic policy? What behavior would you be looking for from a better policy?

13.6 — How does the use of value functions improve the solution?

Problem solving questions

13.7 — Blood management - Part I: Modeling - We are going to consider the blood management problem, but we are going to assume there is only one type of blood, although we are still going to model the aging process, where blood can be 0 to 5 weeks old. Any 5-week old blood that is held must be discarded. As in the book, there are two types of patients: urgent and elective. Let:

$$
\begin{aligned}
R_{t\tau} &= \text{Number of units of blood on hand at time } t \text{ that has been held for } \tau \text{ time periods, } \tau = 0, \ldots, 5.\\
\hat{R}_t &= \text{New blood donations that arrive between } t-1 \text{ and } t, \text{ where } \hat{R}_t \text{ is a scalar.}\\
\hat{D}_t^{urgent} &= \text{New urgent demands that arrive at time } t.\\
\hat{D}_t^{elective} &= \text{New elective demands that arrive at time } t.
\end{aligned}
$$

At time t, we have to decide:

$$
\begin{aligned}
x_t^{urgent} &= \text{Amount of blood to be assigned to urgent patients.}\\
x_t^{elective} &= \text{Amount of blood to be assigned to elective patients.}\\
x_t^{hold} &= \text{Amount of blood to be held.}\\
x_t &= (x_t^{urgent}, x_t^{elective}, x_t^{hold}).
\end{aligned}
$$

Demands do not have to be covered, although the real issue is whether to cover an elective demand now (assuming there is enough blood to cover all the urgent demands) or hold the blood for a potential urgent demand in the future. As before, assume that any demands that are not served leave the system.

Your goal is to maximize a utility function that gives credit of 10 for each urgent patient covered and 5 for each elective patient covered.

a) What is the state variable for this problem?

b) What are the decision variables and exogenous information?

c) What is the transition function?

d) What is the objective function? Assume that we can simulate the policy in a simulator.

e) Create a parameterized cost function approximation that assigns the following costs to each decision:

$$
\begin{aligned}
c^{urgent} &= \text{Penalty for not covering an urgent patient.} \\
c^{elective} &= \text{Penalty for not covering an elective patient.} \\
c^{discard} &= \text{Penalty for discarding blood that exceeds the age of 5 weeks.}
\end{aligned}
$$

As your policy, assume that you are going to minimize these costs at each time period. Treat the vector $c = (c^{urgent}, c^{elective}, c^{discard})$ as a set of tunable parameters. Describe how to optimize the vector c using a stochastic gradient algorithm. Be sure to give the equation for calculating the stochastic gradient.

13.8 — Blood management - Part II: Backward approximate dynamic programming - We are now going to design a policy based on the idea of approximating the value function using backward approximate dynamic programming. This means that you have to specify a linear model to approximate $V_t^x(S_t^x)$. The details of this model are not that important, but you might use something like

$$
\overline{V}_t^x(S_t^x) = \bar{\theta}_{t0} + \sum_{age=0}^{5} \theta_{t1,age} R_{t,age}^{urgent,x} + \sum_{age=0}^{5} \theta_{t2,age} R_{t,age}^{elective,x}. \quad (13.11)
$$

For the purposes of this exercise, you can just write $\overline{V}_t^x(S_t^x) = (\theta_t)^T \phi(S_t^x)$ where θ_t is a column vector of coefficients and $\phi(S_t^x)$ is a column vector of features.

a) Define the post-decision state, and use this to write Bellman's equation to characterize an optimal policy. You will need to write an expression for the value $V_t(S_t)$ of being in the pre-decision state S_t at time t in terms of the value $V_t^x(S_t^x)$ of being in post-decision state S_t^x. You will then need to write an expression for $V_t^x(S_t^x)$ in terms of $V_{t+1}(S_{t+1})$. Assume that units of blood are always integer.

b) What is the dimensionality of the pre- and post-decision state variables? Do we care how large the state space is?

c) Write out detailed pseudo-code that describes how to estimate the value function approximations for this problem over a finite horizon $0, \ldots, T$. Think of this as a programming exercise without the actual programming. It has to be detailed enough that you could hand it to a classmate in the course (and familiar with the material) who could then write the code.

d) Write out the policy using your expression for the approximate value function.

13.9 — Blood management - Part III: Lookahead policy - This time, we are going to assume that we have rolling forecasts of supplies and demands. Let $f^R_{tt'}$ be the forecast of blood donations at time t' made using what we know at time t. Let $f^{D,urgent}_{tt'}$ and $f^{D,elective}_{tt'}$ be the forecasts of new urgent and elective demands arriving at time t' given what we know at time t. Assume that forecasts are provided exogenously (that is, we do not have to model how the forecasts evolve from t to $t+1$). You may use

$$
\begin{aligned}
f^R_t &= (f^R_{tt'})^T_{t'=t+1}, \\
f^{D,urgent}_t &= (f^{D,urgent}_{tt'})^T_{t'=t+1}, \\
f^{D,elective}_t &= (f^{D,elective}_{tt'})^T_{t'=t+1}, \\
f_t &= (f^R_t, f^{D,urgent}_t, f^{D,elective}_t).
\end{aligned}
$$

Let $\sigma^R_{t'-t}$ be the standard deviation of the error between the actual donations $\hat{R}_{tt'}$, which we assume is known based on past performance. We assume that this is purely a function of how far into the future we are planning, given by $t'-t$. Similarly let $\sigma^{D,urgent}_{t'-t}$ and $\sigma^{D,elective}_{t'-t}$ be the standard deviations of the errors in the forecasts of new urgent and elective demands.

a) Model the five elements of a sequential decision problem for this setting. You should be able to copy elements from one of the previous parts to this problem. Feel free to reference any equations by number you wish to re-use. The major change is the inclusion of the forecasts.

b) Write out a DLA policy using a deterministic lookahead with forecasts as point estimates of any future donates and demands.

c) Now design a parameterized policy where you replace each forecast with one that is some number of standard deviations above (or below) the point forecast. Use three parameters, which you might designate $\theta = (\theta^R, \theta^{urgent}, \theta^{elective})$. Write the problem of finding the best value for θ as an optimization problem. Explain any assumptions you have to make in your formulation.

d) Your objective function in part (c) involves approximating an expectation. You can do this via simulation, where you would simulate a sample path ω over a horizon of T time periods. What is meant by ω?

e) Give the formulas for computing the mean and sample variance of the performance of a policy from L simulations using sample paths $\omega^1, \ldots, \omega^L$.

f) Assume that you represent the set of possible values of the vector θ by the sample $\theta^1, \ldots, \theta^K$. Describe a search method using interval estimation parameterized by λ^{IE} (in the book we used θ^{IE}, but this creates too many $\theta's$). You will need to describe your belief model and how it is updated after each time you run a simulation using $\theta = \theta^k$. Assume you have a budget of N simulations, and that λ^{IE} is known.

Programming questions

These exercises use the Python module *BloodManagement.*[1]

13.10 — Our goal is to manage the assignment of different blood types to different patients, who are characterized first by their own blood type, and second by whether the surgery is urgent or elective. This exercise will have you working with two classes of policies: a myopic parametric cost function approximation, and a policy based on value function approximations.

We are going to begin by assuming that you are just going to match different blood types to different demands. Blood is described by blood type (of which there are eight) and age, which will range from 0 to 2 weeks (3 week-old blood is discarded). Patients are described by blood

[1]See http://tinyurl.com/sdagithub

type and whether the surgery is urgent or elective. There are various bonuses and penalties that guide assignments. For example, there are positive bonuses for covering urgent patients (this is highest). There is also a bonus for matching blood types exactly (e.g. A-positive blood with an A-positive patient), and a penalty for discarding blood after it becomes too old.

If we ignore the impact of decisions now on the future, we have a simple linear program that matches supplies and demands, with costs given by this set of bonuses. The problem is that by ignoring the impact of decisions now on the future, we may find that we are not doing the best that we can. One issue that arises is when we use blood now for elective surgery, we are ignoring that this might be useful to hold onto in case we run out of blood for urgent surgery later on. Alternatively, we may use O- blood now rather than hold it for the future when we might run out of other blood types.

Let R_{ta} be the supply of blood with attribute a for week t, and let $R_t = (R_{ta})_{a \in \mathcal{A}}$ where $\mathcal{A}$ is the set of all the different blood attributes (blood type and age). Similarly let D_{tb} be the attributes of a patient where b captures the blood type and whether the surgery is urgent or elective, and let $D_t = (D_{tb})_{b \in \mathcal{B}}$. The state of our system is $S_t = (R_t, D_t)$. Now let $\hat{R}_{t+1,a}$ be the number of units of blood with attribute a that were donated between weeks t and $t+1$. Similarly let $\hat{D}_{t+1,b}$ be the number of new patient arrivals with attribute b. We would write

$$W_{t+1} = (\hat{R}_{t+1,a}, \hat{D}_{t+1,b}).$$

Finally let ω represent a sample path $W_1(\omega), \ldots, W_T(\omega)$ of donations and new patients over our T-week horizon. Assume that we have created a set of simulations of W_t, and let $\Omega = (\omega_1, \ldots, \omega_N)$ be this set of sample realizations.

a) How many dimensions does the state variable S_t have?

b) Let $X^\pi(S_t|\theta)$ be the result of solving the linear program given the state S_t, where θ is the vector of all the bonuses and penalties for different assignments. Let $D_t^{urgent}(x_t)$ be the number of urgent patients that were covered given the decision vector x_t, and let $D_t^{elective}(x_t)$ be the number of elective patients that were covered.

Write the problem of finding the best value of θ as an optimization problem, where instead of our usual expectation you are going to write it as an average over the sample paths in Ω.

c) We are going to consider a dataset where there is a probability that demand occasionally surges. You can set this probability in the spreadsheet. Set this surge probability to 50 percent. There is a special penalty for using blood to cover elective surgery to encourage the myopic model to save blood for urgent surgery that might have an increase in demand later on. Find the best value of this penalty in the set $\{-4, -9, -14, -19, -24\}$ after running 20 testing iterations.

d) Without doing any additional numerical work, imagine now that the penalty on the O-negative blood needs to depend on the week to handle seasonal variations. Since you are simulating 15 weeks, describe a method for optimizing over a 15-dimensional vector (we have described two core strategies in prior assignments - you may pick one, or invent a new one).

13.11 — Now we are going to switch to a VFA-based policy, where we use the marginal value of each blood type (and age) that is held for the future. This will be done with an adaptive learning algorithm that was described in Section 13.4.2 (and very similar to our ADP strategy for the shortest path problem, except now we are doing it for a problem where the decision is a vector).

Set the penalty for using blood on electives to 0. When using a VFA-based policy, the VFA should learn that urgent blood in excess of supply might be needed in the future. When you are using the VFA policy, you will need to run 20 training iterations to estimate the value functions. After these are estimated, you will then run 20 testing iterations to evaluate the quality of the policy.

We are going to test our policies for a dataset where there is a probability demand occasionally surges. You can set this probability in the spreadsheet. Begin by setting this surge probability to 0.7.

a) The adaptive learning algorithm requires estimating the marginal value of each blood type. Let $\hat{v}^n_{ta}$ be our estimate of the marginal value of blood type a for week t while simulating sample path ω^n. Let $\overline{V}_t^{n-1}(R_{ta})$ be our estimate, after $n - 1$ iterations, of the marginal value of the rth unit of blood where $r = R^x_{ta}$ at the end of week t (this is our "post-decision" state variable). Recall that we use $\hat{v}^n_{ta}$ to update the value function approximation around the previous post-decision state variable. We write this updating process as

$$\overline{V}^n_{t-1,a}(R^{x,n}_{t-1,a}) = (1 - \alpha)\overline{V}^{n-1}_{t-1,a}(R^{x,n}_{t-1,a}) + \alpha\hat{v}^n_{ta}.$$

Our first task is that we have to tune α. Run the approximate dynamic programming algorithm for 20 iterations (this is how the spreadsheet is set up) for $\alpha \in \{0, 0.05, 0.1, 0.2, 0.3\}$ and report the results. Note that a stepsize $\alpha = 0$ is the same as keeping the value function approximation equal to zero (in other words, the myopic policy). If $\alpha = 0$, you do not have to train the VFAs, so you just have to run the 20 testing iterations to evaluate the policy.

How well does the VFA policy perform relative to the myopic policy (corresponding to $\alpha = 0$)?

b) Now switch the surge probability to zero, and compare the myopic policy to the VFA policy using a stepsize of $\alpha = 0.2$. How do these compare? Can you explain the behavior for this dataset compared to when there were surges?

14

Optimizing Clinical Trials

14.1 Narrative

At any point in time, pharmaceutical companies may be running hundreds of thousands of clinical trials testing new medications (see clinicaltrials.gov). Drug testing occurs in three phases:

Phase I These are tests run using 20 to 100 volunteers over a few months to determine the dose, identify side effects and perform an initial assessment of drug response and side effects.

Phase II These are larger trials with several hundred patients spanning two years. The goal is to determine if the disease responds to the treatment.

Phase III These are trials involving hundreds to thousands of patients, often spanning multiple years, to assess effectiveness and safety. This is where they determine if the treatment is better than existing treatments.

Both Phase II and Phase III trials require identifying patients with the proper characteristics to enter the trial, at which point they are randomly assigned to one of the groups for comparison.

In our exercise, we assume that we enroll a set of hospitals and clinics each week to achieve a *potential* population, from which patients will be identified as candidates for the trial based on paper records. There is an up-front administrative cost to enroll a hospital or clinic in the trial. The administrative cost reflects the pool of patients that the facility might have for the study. For example, it might cost $250,000 to sign up a group of hospitals and clinics with a total potential population of 500 patients. In our model, we will simply set a $500 per patient signup cost, keeping in mind that this is for a total potential population, from which we draw actual signups.

Once we have signed up a facility, we then advertise the clinical trial from which patients (or their physicians) step forward. At that point, a patient is then subjected to a more detailed evaluation, which determines who is accepted into the trial. Ineligible patients are then dropped.

For the purpose of this exercise, we are going to assume that each patient is given a medication at the beginning of the week. By the end of the week, we know if the patient is responding or not. Patients that respond are designated as successes, the rest as failures. Each week, then, requires an entirely new set of patients, but these are drawn from the base population that we have signed up. We can only increase this population by entering more capacity into the trial, and paying the up-front administrative cost.

14.2 Basic model

We are going to assume that we make decisions at the end of each week t, to be implemented during week $t+1$. We let time t denote the end of week t.

14.2.1 State variables

We have the following state variables:

R_t = The potential population of patients that are in the hospitals and clinics that have been signed up.

α_t = The number of successes for the treatment by week t over the course of the clinical trial.

$$\beta_t \quad = \quad \text{The number of failures for the treatment by week } t.$$
$$\bar{\lambda}_t^{response} \quad = \quad \text{Estimated fraction of potential patients who elect to join the trial given what we know at time } t.$$

Using this information, we can estimate the probability that our treatment is successful using

$$\begin{aligned} \rho_t &= \text{the probability that the treatment is successful given what we know by the end of week } t, \\ &= \frac{\alpha_t}{\alpha_t + \beta_t}. \end{aligned}$$

This means that our state variable would be

$$S^n = (R_t, (\alpha_t, \beta_t), \bar{\lambda}_t^{response}).$$

For the initial state, it is reasonable to assume that $R_0 = 0$, but we may wish to start with an initial estimate of the probability of success (based on prior clinical trials) which we can represent by assuming nonzero values for α_0 and β_0.

14.2.2 Decision variables

We model the number of potential patients that are signed up using:

$$x_t^{enroll} = \text{The increase in the potential population of patients that are acquired by adding new hospital facilities. The number of patients that actually join the clinical trial will be drawn from this population during week } t+1.$$

We also have the decision of when to stop the trial represented by

$$x_t^{trial} = \begin{cases} 1 & \text{continue the trial,} \\ 0 & \text{stop the trial.} \end{cases}$$

If $x_t^{trial} = 0$, then we are going to set $R_{t+1} = 0$, which shuts down the trial. We assume that once we have stopped the trial, we cannot restart it, which means we will require that

$$x_t^{trial} = 0 \quad \text{if } R_t = 0.$$

If we stop the trial, we have to declare whether the drug is a success or a failure,

$$x_t^{drug} = \begin{cases} 1 & \text{if the drug is declared a success,} \\ 0 & \text{if the drug is declared a failure.} \end{cases}$$

We will create policies $X^{\pi^{enroll}}(S_t)$, $X^{\pi^{trial}}(S_t)$ and $X^{\pi^{drug}}(S_t)$ which determine x_t^{enroll}, x_t^{trial} and x_t^{drug}. We can then write

$$X^{\pi}(S_t) = (X^{\pi^{enroll}}(S_t), X^{\pi^{trial}}(S_t), X^{\pi^{drug}}(S_t)).$$

As always, we design the policies later.

14.2.3 Exogenous information

We first identify new patients and patient withdrawals to and from the trial using

$\hat{R}_{t+1}$ = the number of new patients joining the trial during week $t+1$, which depends on the potential population of patients that were signed up, given by $R_{t+1} = R_t + x_t^{enroll}$.

We might, for example, assume that each patient in the population R_{t+1} might sign up for the clinical trial with some probability $\lambda^{response}$ which has to be estimated from data.

We next track our successes with

$\hat{X}_{t+1}$ = the number of successes during week $t+1$,

$\hat{Y}_{t+1}$ = the number of failures during week $t+1$.

The number of failures during week t can be calculated as

$$\hat{Y}_{t+1} = \hat{R}_{t+1} - \hat{X}_{t+1}.$$

These variables depend on the number of patients R_t in the system at the end of week t. As always, we defer to the section on uncertainty modeling the development of the underlying probability models for these random variables.

Our exogenous information process is then

$$W_{t+1} = (\hat{R}_{t+1}, \hat{X}_{t+1}),$$

where we exclude $\hat{Y}_{t+1}$ because it can be computed from the other variables.

14.2.4 Transition function

The transition equation for the number of enrolled patients is given by

$$R_{t+1} = x_t^{trial}(R_t + x_t^{enroll}). \tag{14.1}$$

We update the probability the drug is a success by counting the number of successes and failures using

$$\alpha_{t+1} = \alpha_t + \hat{X}_{t+1}, \tag{14.2}$$
$$\beta_{t+1} = \beta_t + (\hat{R}_{t+1} - \hat{X}_{t+1}). \tag{14.3}$$

Finally, we update our estimate of the number of patients who enroll in the trial by smoothing the current estimate $\bar{\lambda}_t^{response}$ with the latest ratio of the number who enrolled during week $t+1$, $\hat{R}_{t+1}$, and the number who are currently signed up, $R_t + x_t^{enroll}$.

$$\bar{\lambda}_{t+1}^{response} = (1-\eta)\bar{\lambda}_t^{response} + \eta \frac{\hat{R}_{t+1}}{R_t + x_t^{enroll}}. \tag{14.4}$$

Equations (14.1) - (14.4) make up the transition function that we represent generically using

$$S_{t+1} = S^M(S_t, x_t, W_{t+1}).$$

14.2.5 Objective function

We have to consider the following costs:

c^{enroll} = The cost of maintaining a patient in the trial per time period.

c^{trial} = The ongoing administrative overhead costs of keeping the trial going (this stops when we stop testing).

$p^{success}$ = The (large) revenue gained if we stop and declare success, which typically means selling the patent to a manufacturer.

The profit (contribution) in a time period would then be given by

$$C(S_t, x_t) = (1 - x_t^{trial})x_t^{drug}p^{success} - x_t^{trial}(c^{trial} + c^{enroll}x_t^{enroll}). \tag{14.5}$$

Our objective function, then, would be our canonical objective which we state as

$$\max_{\pi} \mathbb{E} \left\{ \sum_{t=0}^{T} C(S_t, X^{\pi}(S_t))|S_0 \right\}, \tag{14.6}$$

where we recognize that our policy is a composition of the patient enrollment policy $X^{\pi^{enroll}}(S_t)$, the trial continuation policy $X^{\pi^{trial}}(S_t)$, and the drug success/failure policy $X^{\pi^{drug}}(S_t)$.

14.3 Modeling uncertainty

There are two potential reasons to develop a formal uncertainty model. The first is for the base model, which we can use both to design policies as well as to run studies. The second is that we may want to model uncertainty in a stochastic lookahead policy.

We begin by developing a probabilistic base model, which means we are going to make a best effort to model the real problem, recognizing that all mathematical models are approximations of the real world.

We need to model three random variables:

- The number of customers $\hat{R}_{t+1}$ that sign up for the trial.
- The (unobservable) success rate ρ^{true}.
- The number of successes $\hat{X}_{t+1}$, which we do observe.

We address each of these below.

14.3.1 The patient enrollment process $\hat{R}_t$

We are going to use the simple model that we make choices (e.g. by signing up hospitals and clinics) that allow us to expect to sign up x_t^{enroll} patients for week $t+1$, giving us a total population of $R_{t+1} = R_t + x_t^{enroll}$. The reality will be different. We propose to model the actual number of arrivals by assuming that they are Poisson with a mean of $\bar{\lambda}^{response}(R_t + x_t^{enroll})$ where $0 < \bar{\lambda}^{response} < 1$ is the fraction of potential patients who elect to join the trial (which is unknown). This means that we can write

$$Prob[\hat{R}_{t+1}(R_t) = r] = \frac{(\bar{\lambda}_t^{response}(R_t + x_t^{enroll}))^r e^{-\bar{\lambda}_t^{response}(R_t + x_t^{enroll})}}{r!}. \quad (14.7)$$

We can use a truncated Poisson distribution for $\hat{R}_{t+1}$, where we have to recognize that the number of patients that join the trial is limited by the number of potential patients given by $R_{t+1} = R_t + x_t^{enroll}$. Let

$$\bar{R}_t = \bar{\lambda}_t^{response}(R_t + x_t^{enroll})$$

be the expected number of patients that will volunteer for the trial (given R_t) and

$$P_{\hat{R}_{t+1}}(r|x_t^{enroll}, \bar{R}_t) = Prob[\hat{R}_{t+1}(x_t^{enroll}) = r|\bar{R}_t].$$

We write $P_{\hat{R}_{t+1}}(r|x_t^{enroll}, \bar{R}_t)$ as a function of x_t^{enroll} and $\bar{R}_t$ to reflect its dependence on the decision and on the number $R_{t+1} = R_t + x_t^{enroll}$.

The truncated Poisson distribution is then given by

$$P_{\hat{R}_{t+1}}(r|x_t^{enroll}, \bar{R}_t) = \begin{cases} \frac{(\bar{R}_t)^r e^{-\bar{R}_t}}{r!}, & r = 0, \ldots, x_t^{enroll} - 1 \\ 1 - \sum_{r=0}^{x_t^{enroll}-1} P_{\hat{R}_{t+1}}(r|x_t^{enroll}, \bar{R}_t) & r = R_t + x_t^{enroll} \end{cases} \quad (14.8)$$

For a population process such as this, a Poisson process is a good starting point. It enjoys the property that the mean equals the variance which equals $\bar{R}_t$.

14.3.2 The success probability ρ^{true}

The successes are driven by the underlying, but unobservable, probability that the treatment will create a success in a patient during a week. We use the Bayesian style of assigning a probability distribution to ρ^{true}. There are three ways to represent the distribution of our belief about ρ^{true}:

- A uniform prior, where we would assume that ρ^{true} is uniformly distributed between 0 and 1.

- A beta distribution with parameters (α_0, β_0).

- A sampled distribution, where we assume that ρ^{true} takes on one of the set of values $(\rho_1, \ldots, \rho_K)$, where we let our initial distribution be

$$p^{\rho}_{0k} = Prob[\rho^{true} = \rho_k].$$

We might let $p_{0k} = 1/K$ (this would be comparable to using the uniform prior). Alternatively, we could estimate these from the beta distribution.

For now, we are going to use our sampled distribution since it is the easiest to work with.

14.3.3 The success process $\hat{X}_t$

The random number of successes $\hat{X}_{t+1}$, given what we know at time t, depends first on the random variable $\hat{R}_{t+1}$ giving the number of patients who entered the trial, and the unknown probability ρ^{true} of success in the trial. The way to create the distribution of $\hat{X}_{t+1}$ is to use the power of conditioning. We assume that $\hat{R}_{t+1} = r$ and that $\rho^{true} = \rho_k$.

Given that r patients enter the trial and assuming that the probability of success is ρ_k, the number of successes $\hat{X}_{t+1}$ is the sum of r Bernoulli (that is, 0/1) random variables. The sum of r Bernoulli random variables is given by a binomial distribution, which means

$$Prob[\hat{X}_{t+1} = s | \hat{R}_{t+1} = r, \rho^{true} = \rho_k] = \binom{r}{s} \rho_k^s (1 - \rho_k)^{r-s}.$$

We can find the unconditional distribution of $\hat{X}_{t+1}$ by just summing over r and k and multiplying by the appropriate probabilities, giving us

$$Prob[\hat{X}_{t+1} = s | \bar{R}_t] = \sum_{k=1}^{K} \left(\sum_{r=0}^{R_t} Prob[\hat{X}_{t+1} = s | \hat{R}_{t+1} = r, \rho^{true} = \rho_k] P_{\hat{R}_{t+1}}(r | x_t^{enroll}, \bar{R}_t) \right) p^{\rho}_{tk}. \tag{14.9}$$

Using explicit probability distributions such as the one for $\hat{X}_{t+1}$ in Equation (14.9) is nice when we can find (and compute) them, but there are many complex problems where this is not possible. For example, even Equation (14.9) required that we use the trick of using a sampled representation of the continuous random variable ρ^{true}. Without this, we would have had to introduce an integral over the density for ρ^{true}.

Another approach, which is much easier and extends to even more complicated situations, uses Monte Carlo sampling to generate $\hat{R}_{t+1}$ and $\hat{X}_{t+1}$. This process is outlined in Figure 14.1, which produces a sample $\hat{X}^1_{t+1}, \ldots, \hat{X}^N_{t+1}$ (and corresponding $\hat{R}^1_{t+1}, \ldots, \hat{R}^N_{t+1}$). We can now approximate the random variable $\hat{X}_{t+1}$ with the set of outcomes $\hat{X}^1_{t+1}, \ldots, \hat{X}^N_{t+1}$, each of which may occur with equal probability.

Step 1. Loop over iterations $n = 1, \ldots, N$:

- **Step 2a.** Generate a Monte Carlo sample $r^n \sim \hat{R}_{t+1}(x^{enroll})$ from the Poisson distribution given by Equation (14.8).
- **Step 2b.** Generate a Monte Carlo sample of the true success probability $\rho^n \sim \rho^{true}$.
- **Step 2c.** Given r^n and ρ^n, loop over our r^n patients and generate a 0/1 random variable which is 1 (that is, the drug was a success) with probability ρ^n.
- **Step 2d.** Sum the successes and let this be a sample realization of $\hat{X}^n_{t+1}$.

Step 3. Output the sample $\hat{X}^1_{t+1}, \ldots, \hat{X}^N_{t+1}$.

Figure 14.1: A Monte Carlo-based model of the clinical trial process.

14.4 Designing policies

We are going to use this problem to really understand our full stochastic lookahead policy which we first introduced in Chapter 7, given by

$$X^*(S_t) = \arg\max_{x_t \in \mathcal{X}} \Bigg(C(S_t, x_t) +$$
$$\mathbb{E}_{W_{t+1}} \left\{ \max_{\pi} \mathbb{E}_{W_{t+2},\ldots,W_T} \left\{ \sum_{t'=t+1}^{T} C(S_{t'}, X^{\pi}_{t'}(S_{t'})) | S_{t+1} \right\} | S_t, x_t \right\} \Bigg). \tag{14.10}$$

We are going to focus on what is meant by that maximization over policies π embedded within the policy (this might be called the "policy-within-the-policy."

For our clinical trials application, we have to design policies for the three different decisions: the number of patients to enroll, whether or not to continue the trial, and whether or not the drug is declared a success when the trial is stopped. We are going to begin by designing simple policy function approximations for the decisions of whether to stop or continue, and if we stop, whether we declare the drug a success or a failure. We then address the more difficult decision of how many patients to enroll in the trial.

14.4.1 Stopping the trial and declaring success or failure

We begin by using our belief about ρ^{true} given by the beta distribution with parameters (α_t, β_t), which gives us an estimate of

$$\bar{\rho}_t = \frac{\alpha_t}{\alpha_t + \beta_t}.$$

Now introduce the parameters $\theta^{stop-low}$ and $\theta^{stop-high}$, where we are going to stop the trial and declare success if $\bar{\rho}_t > \theta^{stop-high}$, while we will stop the trial and declare failure if $\bar{\rho}_t < \theta^{stop-low}$. Let $\theta^{stop} = (\theta^{stop-low}, \theta^{stop-high})$. We use these rules to define the policy for stopping the trial as

$$X^{trial}_t(S_t|\theta^{stop}) = \begin{cases} 1 & \text{if } \theta^{stop-low} \leq \bar{\rho}_t \leq \theta^{stop-high}, \\ 0 & \text{otherwise.} \end{cases}$$

If we stop the trial, then the policy for declaring success (1) or failure (0) is given by

$$X^{drug}_t(S_t|\theta^{stop}) = \begin{cases} 1 & \text{if } \bar{\rho}_t > \theta^{stop-high}, \\ 0 & \text{if } \bar{\rho}_t < \theta^{stop-low}. \end{cases}$$

14.4.2 The patient enrollment policy

It is often the case that problems with a physical state (such as R_t) need a lookahead policy, just as we used with our stochastic shortest path problem. But, as we saw with the stochastic shortest path problem, we get to choose what to put in our stochastic lookahead model.

One choice we have to make is the stopping policy $X^{trial}(S_t|\theta^{stop})$ and the success/failure policy $X^{drug}(S_t|\theta^{stop})$, where we propose to use the same parameter vector θ^{stop} in our lookahead model as we do in our base model. We can refer to these as $\tilde{X}^{trial}(\tilde{S}_t|\theta^{stop})$ and $\tilde{X}^{drug}(\tilde{S}_t|\theta^{stop})$, since these now apply only to the lookahead model.

The problem of determining how many new potential patients to enroll is somewhat more difficult, since it is necessary to pay an upfront cost to acquire more potential patients, and we have to do this under uncertainty about the willingness of patients to join the trial (given by the unknown parameter $\lambda^{response}$).

To create a full lookahead model as we described in Section 14.3, we would create variables such as $\tilde{\lambda}_{tt'}$ for the lookahead version of $\bar{\lambda}_t^{response}$, $\tilde{\rho}_{tt'}$ for $\bar{\rho}_t$, and $(\tilde{\alpha}_{tt'}, \tilde{\beta}_{tt'})$ for (α_t, β_t). Otherwise, all the logic would be the same as the original uncertainty model.

While we can use the full uncertainty model, we can choose to simplify the model in different ways. These choices include:

- The enrollment rate $\bar{\lambda}_t^{response}$ - We have two options:
  - We can continue to estimate $\bar{\lambda}_t^{response}$, where we would introduce the notation $\tilde{\lambda}_{tt'}^{response}$ as the estimate at time t' in the lookahead model of the enrollment rate λ.
  - We could fix $\tilde{\lambda}_{tt'}^{response} = \bar{\lambda}_t^{response}$, which is our estimate at time t in the base model.
- The drug success rate ρ^{true} - We again have two options:
  - We can continue to estimate the success rate. For this, we would define the variables $(\tilde{\alpha}_{tt'}, \tilde{\beta}_{tt'})$ for accumulating successes and failures in the lookahead model.

– Alternatively we could fix $(\tilde{\alpha}_{tt'}, \tilde{\beta}_{tt'}) = (\alpha_t, \beta_t)$ within the lookahead model.

Using our choices for modeling uncertainty, we can suggest three different strategies for designing a lookahead model:

Model A Deterministic lookahead model - Here, we are going to assume that the enrollment rate $\tilde{\lambda}^{response}_{tt'} = \bar{\lambda}^{response}_t$, which means that the enrollment rate is fixed at the estimate at time t when we create the lookahead model. We then assume that the true drug success probability is fixed at

$$\tilde{\rho}_{tt'} = \bar{\rho}_t = \frac{\alpha_t}{\alpha_t + \beta_t},$$

which is our estimate at time t in the base model.

Model B We fix our estimate of the enrollment rate at $\tilde{\lambda}_{tt'} = \bar{\lambda}^{response}_t$, but assume that we continue learning about the effectiveness of the drug.

Model C We model the process of learning the enrollment rate $\tilde{\lambda}_{tt'}$ and the drug effectiveness $\tilde{\rho}_{tt'}$.

Note that we did not include the potential fourth model where we fix the drug effectiveness but continue learning the patient enrollment rate (we will see in a minute how silly this model would be).

We are going to use these three models to illustrate the process of designing a lookahead model.

14.4.3 Model A

Model A is a deterministic problem, since we are fixing both the estimated enrollment rate $\tilde{\lambda}_{tt'} = \bar{\lambda}^{response}_t$, and $\tilde{\rho}_{tt'} = \bar{\rho}_t$. The good news is that this is basically a deterministic shortest path problem, where the number of patients we have signed up (in the lookahead model), given by $\tilde{R}_{tt'}$, is like a node in a network, and the decision $\tilde{x}^{enroll}_{tt'}$ is a link that takes us to node $\tilde{R}_{t,t'+1} = \tilde{R}_{tt'} + \tilde{x}^{enroll}_{tt'}$.

To see this, recall Equation (5.1) for our deterministic shortest path problem, which we repeat here

$$v_i = \min_{j \in \mathcal{N}_i^+} (c_{ij} + v_j).$$

Now we just replace v_i for the value at node i, with $\tilde{V}_{tt'}(\tilde{R}_{tt'})$ which is the value of having $\tilde{R}_{tt'}$ patients signed up (remember we are in our lookahead model). The decision to go to node j is replaced with the decision to sign up $\tilde{x}_{tt'}^{enroll}$ patients. Instead of this taking us to node j, it takes us to node $\tilde{R}_{tt'} + \tilde{x}_{tt'}^{enroll}$. So Bellman's equation becomes

$$\tilde{V}_{tt'}(\tilde{R}_{tt'}) = \min_{\tilde{x}_{tt'}^{enroll}} \left(\tilde{C}(\tilde{R}_{tt'}, \tilde{x}_{tt'}^{enroll}) + \tilde{V}_{t,t'+1}(\tilde{R}_{tt'} + \tilde{x}_{tt'}^{enroll})\right). \quad (14.11)$$

The one-period profit function $\tilde{C}(\tilde{R}_{tt'}, \tilde{x}_{tt'}^{enroll})$ is adapted from the same function for our base model (see Equation (14.5)).

There is only one problem with our deterministic lookahead model: we would never stop, because our policy for stopping requires that our estimate of $\tilde{\rho}_{tt'}$ moves into the "success" or "fail" regions (it would have to start in the "continue" region, since otherwise we would have stopped the base model). However, this does not mean that we cannot use the deterministic lookahead model: we just have to fix a horizon H and stop when $t' = t + H$.

Using this strategy, we solve our deterministic shortest path problem over the horizon $t' = t, \ldots, t + H$, and then from this find $\tilde{x}_{tt}^*$. Our enrollment policy is then

$$X^{\pi^{enroll}}(S_t) = \tilde{x}_{tt}^*.$$

We are not claiming that this will be an effective policy. We are primarily illustrating the types of modeling approximations that can be made in a lookahead model.

14.4.4 Model B

Now we are going to fix our estimate of the response rate $\tilde{\lambda}_{tt'}$ at our estimate $\bar{\lambda}_t^{response}$ at time t in the base model. To simplify our model, we are going to assume that the number of enrollments $\tilde{\hat{R}}_{t,t'+1}$ equals the expected number of patients that will volunteer $\bar{\tilde{R}}_{tt'}$. The enrollments $\hat{\tilde{R}}_{t,t'+1}$ are generated deterministically from

$$\hat{\tilde{R}}_{t,t'+1} = \lfloor \bar{\lambda}_t^{response}(\tilde{R}_{tt'} + \tilde{x}_{tt'}^{enroll}) \rfloor,$$

where $\lfloor x \rfloor$ means to round x down to the nearest integer. We then compute the distribution of $\hat{\tilde{X}}_{t,t'+1}$ using $Prob[\hat{X}_{t+1} = s|\bar{R}_t]$ but where we replace $\bar{R}_t$ with $\bar{\tilde{R}}_{tt'}$.

We still have to generate the number of successes $\hat{\tilde{X}}_{t,t'+1}$ from a simulated truth $\tilde{\rho}_{tt'}$, from which we will update $(\tilde{\alpha}_{tt'}, \tilde{\beta}_{tt'})$ which we do using

$$\begin{aligned}
\tilde{\alpha}_{t,t'+1} &= \tilde{\alpha}_{tt'} + \hat{\tilde{X}}_{t,t'+1}, \\
\tilde{\beta}_{t,t'+1} &= \tilde{\beta}_{tt'} + \bar{\tilde{R}}_{tt'} - \hat{\tilde{X}}_{t,t'+1}.
\end{aligned}$$

We model the distribution of $\hat{\tilde{X}}_{t,t'+1}$ using $Prob[\hat{\tilde{X}}_{t,t'+1} = s|\bar{\tilde{R}}_{tt'}]$ in Equation (14.9) but conditioning on $\bar{\tilde{R}}_{tt'}$ instead of $\bar{R}_t$ (remember that we can also use the sampled distribution using the method in Figure 14.1 instead of the Poisson distribution).

We can solve the lookahead model by adapting Bellman's equation for Model A in Equation (14.11) for $t' = t, \ldots, t+H$:

$$\begin{aligned}
\tilde{V}_{tt'}(\tilde{S}_{tt'}) &= \min_{\tilde{x}_{tt'}^{enroll}} \Bigg(\tilde{C}(\tilde{S}_{tt'}, \tilde{x}_{tt'}^{enroll}) + \\
&\quad \sum_{s=0}^{\bar{\tilde{R}}_{tt'}} Prob[\hat{\tilde{X}}_{t,t'+1} = s|\bar{\tilde{R}}_{tt'}] \tilde{V}_{t,t'+1}(\tilde{S}_{t,t'+1}|\hat{\tilde{X}}_{t,t'+1} = s) \Bigg),
\end{aligned}$$

where $\tilde{S}_{t,t'+1} = (\tilde{R}_{t,t'+1}, \tilde{\alpha}_{t,t'+1})$ is conditioned on the number of successes $\hat{\tilde{X}}_{t,t'+1} = s$, and where $Prob[\hat{\tilde{X}}_{t,t'+1} = s|\bar{\tilde{R}}_{tt'}]$ comes from Equation (14.9). We have to keep in mind that the evolution of $\tilde{R}_{tt'}$ has to reflect if we have decided to stop or continue the trial within the lookahead model.

Our physical state variable (total potential patients) $\tilde{R}_{t,t'+1}$ is given by

$$\tilde{R}_{t,t'+1} = \begin{cases} \tilde{R}_{tt'} + \tilde{x}_{tt'}^{enroll} & \text{if } \tilde{X}^{trial}(\tilde{S}_{tt'}|\theta^{stop}) = 1, \\ 0 & \text{otherwise.} \end{cases}$$

Note that as with our base model, the number of potential patients drops to zero if we stop the trial in the lookahead model.

Our current estimate of the success of the drug (in the lookahead model) is computed using

$$\bar{\tilde{\rho}}_{tt'} = \frac{\tilde{\alpha}_{tt'}}{\tilde{\alpha}_{tt'} + \tilde{\beta}_{tt'}}.$$

In the expectation, if we condition on the number of successes being $\hat{\tilde{X}}_{t,t'+1} = s$, then the updated belief state $(\tilde{\alpha}_{tt'}, \tilde{\beta}_{tt'})$ is

$$\tilde{\alpha}_{t,t'+1} = \tilde{\alpha}_{tt'} + s, \tag{14.13}$$

$$\tilde{\beta}_{t,t'+1} = \tilde{\beta}_{tt'} + (\bar{\tilde{R}}_{tt'} - s). \tag{14.14}$$

We now have to solve the lookahead model using Bellman's equation in Equation (14.12). For this problem, it makes sense to use a large-enough horizon H so that we can confidently assume that we would have stopped the trial by then (that is $\tilde{X}^{trial}(\tilde{S}_{tt'}|\theta^{stop}) = 0$). This means we can assume that $\tilde{V}_{t,t+H}(\tilde{S}_{t,t+H}) = 0$, and work backward from there to time t. Once we have solved the dynamic program, we can pull out our enrollment decision using

$$X^{enroll}_t(S_t) = \arg\min_{\tilde{x}^{enroll}_{tt}} \left(\tilde{C}(\tilde{S}_{tt}, \tilde{x}^{enroll}_{tt}) + \sum_{s=0}^{\bar{\tilde{R}}_{t,t+1}} Prob[\hat{\tilde{X}}_{t,t'+1} = s|\bar{\tilde{R}}_{tt}]\tilde{V}_{t,t+1}(\tilde{S}_{t,t+1}|\hat{\tilde{X}}_{t,t'+1} = s) \right).$$

14.4.5 Model C

Model C models the process of learning the enrollment rate $\tilde{\lambda}_{tt'}$ and the drug effectiveness $\tilde{\rho}_{tt'}$.

Model C is almost the same as the base model, since we are modeling all the different forms of uncertainty. The only way that it is a lookahead model would be our introduction of the simplified policy (our "policy function approximation") for stopping the trial, and for determining if the drug is a success. However, we could ignore these policies and formulate the entire problem as a dynamic program, using the full state variable.

14.5 What did we learn?

- We introduce the clinical trial problem to illustrate the challenges of a problem that involves active learning (there are belief state variables) while also managing a finite resource (the budget for testing patients).
- We illustrate three types of uncertainty: the patient enrollment process, the probability that the drug is successful, and the process of successes for individual patients.
- We identify two decisions: whether to stop the trial and declare success or failure, and the admission of patients into the trial.
- We then design three types of lookahead policies that are distinguished by how we approximate the lookahead model, and how we approximate policies in the lookahead model.

14.6 Exercises

Review questions

14.1 — The state variable S^n includes beliefs about two uncertain quantities. What are these uncertain quantities, and how are beliefs about them captured in the state variable?

14.2 — Explain the three decisions that have to be made during the clinical trial.

14.3 — Explain the types of exogenous information, and how they are affected by decisions and the state of the system.

14.4 — Explain the logic of the lookahead policy called Model A, and discuss its strengths and weaknesses.

14.5 — Explain the logic of the lookahead policy called Model B, and discuss its strengths and weaknesses.

14.6 — Explain the logic of the lookahead policy called Model C, and discuss its strengths and weaknesses.

Problem solving questions

14.7 — In Section 14.4, we wrote the full direct lookahead policy as

$$X^*(S_t) = \arg\max_{x_t \in \mathcal{X}} \Bigg(C(S_t, x_t) + \\ \mathbb{E}_{W_{t+1}} \left\{ \max_{\pi} \mathbb{E}_{W_{t+2},\ldots,W_T} \left\{ \sum_{t'=t+1}^{T} C(S_{t'}, X_{t'}^{\pi}(S_{t'})) | S_{t+1} \right\} | S_t, x_t \right\} \Bigg). \tag{14.15}$$

If we could actually compute this, we would have an optimal policy. We are going to explore this policy, and then apply it to our clinical trial problem.

a) Assume that each random variable $W_{t+1}, \ldots, W_T$ can only take outcomes 0 or 1. Next assume that the decision $x_t, \ldots, x_T$ can also take on just the values 0 or 1. The policy in Equation (14.15) can be illustrated as a decision tree. Draw the tree for the horizon $t, t+1, t+2$.

b) What is the structure of the policy $X_{t'}^{\pi}(S_{t'})$ represented by the decision tree in part (a)? Said differently, what type of policy is $X_{t'}^{\pi}(S_{t'})$ when given by a decision tree?

14.8 — Since we generally cannot compute Equation (14.15), we have to replace the full lookahead model with an approximate lookahead model that we write as

$$X^{DLA}(S_t) = \arg\min_{x_t \in \mathcal{X}} \Bigg(C(S_t, x_t) + \\ \tilde{E}_{\widetilde{W}_{t,t+1}} \left\{ \min_{\tilde{\pi}} \mathbb{E}_{\widetilde{W}_{t,t+2},\ldots,\widetilde{W}_{tT}} \left\{ \sum_{t'=t+1}^{T} C(\tilde{S}_{tt'}, \tilde{X}_t^{\tilde{\pi}}(\tilde{S}_{tt'})) | \tilde{S}_{t,t+1} \right\} | S_t, x_t \right\} \Bigg). \tag{14.16}$$

where the dynamics of our approximate lookahead model are governed by

$$\tilde{S}_{t,t'+1} = S^M(\tilde{S}_{tt'}, X_{t'}^{\tilde{\pi}}(\tilde{S}_{tt'}), \widetilde{W}_{t,t'+1}). \tag{14.17}$$

a) Imagine that our policy in the lookahead model is a parametric function such as "stop enrolling patients when $\bar{\rho}_t$ falls outside of the range $[\tilde{\theta}^{stop-low}, \tilde{\theta}^{stop-high}]$." For the purpose of this question, we can also replace the enrollment policy with a simple "enroll $\tilde{\theta}^{enroll}$ patients" (this would be a static parameter). We can write this function as $X^{\tilde{\pi}}(\tilde{S}_{tt'})$ where $\tilde{\theta} = (\tilde{\theta}^{stop-low}, \tilde{\theta}^{stop-high}, \tilde{\theta}^{enroll})$. How would you rewrite Equation (14.16) to reflect that the policy in the lookahead model is a parametric function?

b) Part (a) implies that we have to find the best $\tilde{\theta}$ for a given state $\tilde{S}_{t,t+1}$ at time t. This means that the optimal solution is actually a function $\tilde{\theta}_{t+1}(\tilde{S}_{t,t+1})$. This would have to be computed given that we are in the simulated state $\tilde{S}_{t,t+1}$ in the approximate lookahead model.

   In practice, finding the optimal policy each time we step forward seems complicated (and expensive), since we would have to stop and tune $\tilde{\theta}$ for any state $\tilde{S}_{t,t+1}$ that we land in.

   Imagine now that we would like to simplify the process by finding just one θ that we use for all times t, and any state $\tilde{S}_{t,t+1}$. Write the optimization problem that you would have to solve to find this value of θ.

c) What changes in Equations (14.16) and (14.17) if we replace the random variable $\widetilde{W}_{t,t'+1}$ in the lookahead model with a point forecast $f^W_{tt'}$? Continue to assume that the policy is the parametric function we introduced in part (a).

d) Describe the approximations made in the lookahead model B in Section 14.4.4 for the clinical trial problem.

Programming questions

These exercises use the Python module *ClinicalTrialsDriverScript.py*.[1]

14.9 — Set the trial size to $T = 50$, the lookahead horizon to $H = 5$ and run a simulation of Model A. Record the stopping time and explain why the deterministic lookahead model yields the same number of new potential patients x_t^{enroll} at each time t.

[1] See http://tinyurl.com/sdagithub

14.10 — Now set the lookahead horizon to $H = 50$. Modify the module *ClinicalTrialsDriverScript.py* to include a for-loop and run 10 simulations (testing iterations) of Model B. Compute the average revenue over all simulations.

14.11 — When choosing $\theta^{stop} = (\theta^{stop-low}, \theta^{stop-high})$ for our PFA for determining when to stop, we usually choose a large enough $\theta^{stop-high}$ to make sure the drug is successful. Conversely, we choose a large $\theta^{stop-low}$ so that, if the true success rate of the drug is low, we stop the trial early before we lose too much money. However, we cannot make $\theta^{stop-low}$ too high, or else we risk stopping the trial before we have enough information about the drug's true success rate.
Fix $\theta^{stop-high} = 0.8$ and vary $\theta^{stop-low}$ in the interval [0.77, 0.79], in increments of 0.005. For each resulting $(\theta^{stop-low}, \theta^{stop-high})$, run 5 simulations of Model B and compute the average revenue. Plot the resulting revenues against the values of $\theta^{sto-low}$.

14.12 — Models A and B each solve a lookahead problem in which at least one of the estimates $\bar{\lambda}_{tt'}$ and $\bar{p}_{tt'}$ is fixed at time t in the base model. Model C uses only the base model in the form of a hybrid policy search-VFA policy to model the process of learning both the enrollment rate $\bar{\lambda}_{tt'}$ and $\bar{p}_{tt'}$. However, we can create a lookahead version of Model C (called Model C Extension) in which the enrollments $\tilde{\hat{R}}_{t,t'+1}$ are generated from the truncated Poisson distribution with mean

$$\tilde{\hat{R}}_{t,t'+1} = [\bar{\lambda}_t^{response}(\tilde{R}_{tt'} + \tilde{x}_{tt'}^{enroll})]$$

and the distribution of $\tilde{\hat{X}}_{t,t'+1}$ is the same as in Model B.
Your task is to implement the Model C Extension by adding the method *model_C_extension_value_fn* to the Python module *ClinicalTrialsPolicy.py*. The method uses the model_C_extension_policy (which is already in the code) calls the *model_C_extension_value_fn* to compute the value function for the Bellman equation. To write the method *model_C_extension_value_fn*, copy the code from *model_B_value_fn* and add an additional for-loop for the new enrollments in $[0; x^{enroll})$ in steps of $x^{enroll}/10$. Modify the step value and Bellman cost to account for the Model C Extension (hint: use the *trunc_probs* method). Set

the trial size to $T = 50$, the lookahead horizon to $H = 5$ and run one simulation of Model C Extension. Report the stopping time and the revenue.

Acknowledgments

Any proper acknowledgment of the work behind this book would recognize everyone who contributed to the graduate-level text, *Reinforcement Learning and Stochastic Optimization: A unified framework for sequential decisions*. There are simply too many people to list them all here, and I ask readers to check the Acknowledgments section in that book for my best effort at recognizing the efforts of so many who contributed to my understanding of sequential decision problems.

This said, I would like to recognize a few people who contributed to this book. First there was an enthusiastic group of interns who wrote all the Python modules that are used in the exercises for this book: Raluca Cobzaru, Andrei Grauer, Joy Hii, John Nguyen and Robert Raveaunu.

Second I warmly acknowledge the efforts of Dr. Juliana Nascimento, who went through every line of this Python code, fixing bugs, cleaning the logic, and helping me write the problem sets that were based on these exercises.

Finally and most important was my undergraduate class, ORF 411: Sequential Decision Analytics and Modeling, that signed up for the course and participated in the first course specifically on "sequential decision analytics" taught anywhere. They helped me refine the lectures which can be found at https://tinyurl.com/RLSOcourses/ (scroll down to "Undergraduate/masters course in sequential decision analytics" for the slides).[2]

Warren B. Powell
Princeton, New Jersey
August, 2022

[2]The lectures are also available for download via http://dx.doi.org/10.1561/0200000103_supp

References

Powell, W. B. (2022). *Reinforcement Learning and Stochastic Optimization: A unified framework for sequential decisions.* Princeton NJ: John Wiley and Sons. 1–1100. URL: http://www.castlelab.princeton.edu/RLSO/.

Spall, J. C. (2003). *Introduction to Stochastic Search and Optimization: Estimation, simulation and control.* Hoboken, NJ: John Wiley and Sons.

Sterman, J. D. (1989). "Modeling Management Behavior - Misperceptions of Feedback in a Dynamic Decision-Making Experiment". *Management Science.* 35: 321–339.

Tversky, A. and D. Kahneman. (1974). "Judgment under uncertainty: Hueristics and biases". *Science.* 185: 1124–1131.

www.ingramcontent.com/pod-product-compliance
Lightning Source LLC
LaVergne TN
LVHW051051220926
848485LV00001BA/72

* 9 7 8 1 6 3 8 2 8 0 8 2 8 *